EMERGENCE OF THE OMNIVERSE

all Universes plus the spiritual realm

Emergence of the Omniverse

Universe - Multiverse - Omniverse

Accessing Spiritual Freedom in the Omniverse on New Earth
by Letting Go of the Transhumanist Agenda,
AI Artificial Intelligence Singularity, and
False Afterlife/Interlife Reincarnation Matrix

By Alfred Lambremont Webre, JD, MEd

Author of *Exopolitics: Politics, Government, and Law in the Universe*

UniverseBooks.com

Emergence of the Omniverse

Emergence of the Omnivers

Universe - Multiverse - Omniverse

Accessing Spiritual Freedom in the Omniverse on New Earth
by Letting Go of the Transhumanist Agenda,
AI Artificial Intelligence Singularity, and
False Afterlife/Interlife Reincarnation Matrix

ALFRED LAMBREMONT WEBRE, J.D., M.Ed.

AUTHOR OF *EXOPOLITICS: POLITICS, GOVERNMENT, AND LAW IN THE UNIVERSE*

Library and Archives Canada Cataloguing in Publication
Webre, Alfred Lambremont, 1942 -

ISBN 978-0-9737663-4-9
Covered by MiblArt

Edited by Professor Bob Stannard
www.twilocity.com

Spanish language Edition Translation by Francia Vega Group
www.studio-traduction.com

"With deep thanks and gratitude to Francia Vega and a dedicated Team of Translators who labour with care, Love, and skill to bring this knowledge of the Omniverse - the third great cosmological body after the Universe and the Multiverse through which humanity comprehends the Cosmos - to the awareness and understanding of the Spanish speaking world and other languages. In Light and Life, Alfred Lambremont Webre, Vancouver, British Columbia, Canada January 6, 2020."

1. Omniverse 2. Multiverse 3. Universe 4. God 5. Soul 6. Spirit 7. Afterlife 8. Reincarnation 9. Intelligent life 10. Extraterrestrial 11. Science 12. Exopolitics

BOOKS BY ALFRED LAMBREMONT WEBRE:

Emergence of the Omniverse

(Vancouver, Canada: Universebooks.com, 2020)

English; Spanish; French; Chinese Editions

The Omniverse: Transdimensional Intelligence, Time Travel, the Afterlife, and the Secret Colony on Mars

(Rochester, Vermont: Bear & Co./Inner Traditions Dec. 2015)

English; Chinese; German; Hungarian Editions

My Journey Landing Heaven on Earth

(Vancouver, Canada: Universebooks.com, 2015)

The Dimensional Ecology of the Omniverse

(Vancouver, Canada: Universebooks.com, 2014)

Exopolitica: La Política, El Gobierno y La Ley en el Universo

(Granada, Spain: VesicaPiscis.eu 2009)

Spanish Version

Exopolitics: Politics, Government and Law in the Universe

(Vancouver, Canada: Universebooks.com, 2005)

The Levesque Cases

(Ontario: PSP Books, 1990)

The Age of Cataclysm

(New York: G.P. Putnam's Sons, 1974);

(New York: Berkeley Medallion, 1975); (Capricorn Books, 1975);

Japanese version (Tokyo: Ugaku Sha, 1975)

Emergence of the Omniverse

Andrew D. Basiago, JD, MPhil (Cantab), Former U.S. chrononaut—time traveler—in the DARPA Project Pegasus Secret Time Travel program as well as a U.S. Mars Astronaut, states,

> I know that the Afterlife and reincarnation is not just theoretical but it is real! I can remember many of my deaths. I have met eleven [11] "deceased" people who are in the Afterlife, including my father Raymond F. Basiago.
>
> As part of my training in DARPA-CIA's Project Pegasus secret U.S. government time travel program, I was regressed to a prior life as a geometry professor at a New England college—possibly Yale, Amherst, or Boudoin—where I experienced firsthand that the soul cannot die.
>
> The woman scientist who regressed me stated, "DARPA had proved that the soul does not die; we are regressing you to a prior life so in case you die during time travel, you understand your soul will not die."
>
> There are differing Afterlife conditions whereby a soul might be Earth-bound after one death in order to reincarnate and might return to the Godhead after another death.
>
> Typically, individuals like you and I might have up to 10,000 reincarnations on Earth before we are ready to ascend. (Basiago & Webre, 2019)

Wes Penre (2018), researcher and writer, researching metaphysics, extraterrestrial intervention with Earth, spiritual growth, and Artificial Intelligence, writes,

> In the "Wes Penre Papers," as well as in my e-book, *Synthetic Super-Intelligence and the Transmutation of Humankind—A Roadmap to the Singularity*, I wrote about what normally happens after a person dies and leaves his or her third-dimensional human body. I argued that the "afterlife" is just another aspect of the frequency prison in which we are contained. The afterlife, I wrote, is just an intermediary station where souls can rest, until they are recycled (reincarnated) into a new life on Earth. I also suggested what we could do to avoid this afterlife trap and break out of the frequency prison, in which we are trapped as if we were criminals when we didn't commit any crime in the first place. Because of the importance of this information, I am writing this article in an effort to summarize these concepts and to make it more available in one place.
>
> As spiritual beings, we were seduced and manipulated into entering these physical and dense vessels that we are calling "human" bodies a very long time ago, and by doing so, we were trapped in this limited frequency band we call the Third Dimension (see the above references to other parts of my work for more details). Once locked into this frequency, we got stuck in a hologram created by an extraterrestrial group, which most people currently call the "Anunnaki," but whom I will call the "Alien Invader Force" (AIF) or the "Overlords" in this article. This ET group is controlled by a being ancient Sumerian texts call En.ki (who is also the entity

behind the Lucifer archetype). This hologram consists of everything we can perceive with our five senses, and therefore includes the entire known universe.

It would be correct to say that everything that is made of atoms is part of the hologram, which is a distortion and is only a small part of what would be considered the real Universe. (paras. 1-3)

Table of Contents

Page

Introduction 22

The Transhumanist Agenda and the Afterlife/Interlife Matrix 23

Chapter One: My Introduction to the Dimensional Ecology of the Omniverse 27

My Personal Story 28

Context Communication Theory of Extraterrestrial Communications 30

The 1977 Carter White House Extraterrestrial Communication Study 30

Earth Versus the Shadow Government, the Deep State, and the Simulacrum 32

Prevention and Coopting of Evolutionary Input from the Dimensional Ecology of the Omniverse to the Simulacrum Earth 32

The UFO/ET "Adductions" [Abductions] of Jimmy Carter, Alfred Lambremont Webre, and Andrew D. Basiago 33

The 2017 Revelations of the Jimmy Carter, Alfred Lambremont Webre, and Andrew D. Basiago ET "Adduction" 34

Basiago: "The CIA betrayed Carter because he was an ET Abductee" 34

Ethical ETs "Adduct" President Jimmy Carter, Alfred Lambremont Webre, and Chrononaut Andrew D. Basiago 1969-75 35

Quantum Entanglement Compared to "Adduction" Entanglement 36

Who are the ETs behind the Carter-Webre-Basiago Adductions? 38

Exopolitical Impacts of the Carter, Webre, and Basiago ET "Adduction" 39

ET "Adductee" Andrew D. Basiago: Truth, Reform, Innovation, and the Advent of Global Teleportation 42

Three Pillars: Truth, Reform, and Innovation. 42

How to Support the Missions of ET "Adductees" President Jimmy Carter, Alfred Lambremont Webre, and Andrew D. Basiago 46

Chapter Two Universe - Multiverse - Omniverse 49

Leading Discoveries From the Omniverse FAQ: Frequently Asked Questions 49

Wes Penre's Analysis of the Forced Birth-Death Cycle, the Afterlife/Interlife Matrix, and Their Implications for Soul Liberation From the AI Singularity at Bodily Death 58

Chapter Three Emergence of the Omniverse 67
2010: Omnisense and the Omniverse 67
Omniverse 67
2011: Ethan Zaghmut Wise and the Omniverse 67
2014: Scientist David Bertolacci—The Omniverse and Scientific, Pantheist, Non-Theist and Atheist Perspectives 72
2014: The Omniverse—Wes Penre and the KHAA 75
2016: A New Dawn for the Omniverse 79

Chapter Four Science and The Spiritual Dimensions of the Omniverse 95
Parapsychology and the Science of the Omniverse 96
The Science of the Omniverse: Omniverse Equation 97
Source (God) in the Omniverse 98
The Dimensional Ecology of the Omniverse Hypothesis 100
The Dimensional Ecology of the Omniverse: The Divine Female as Source of KHAA 100

Chapter Five Near Death Experience [NDE] as Evidence of the Afterlife/Interlife Matrix 107
Can Experiences Near Death Furnish Evidence of Life After Death? 109
Deathbed Visions 118
Eben Alexander, MD 121

Chapter Six Out of Body Experiences (OOBEs), Instrumental Transcommunication (ITC), Chronovision: Evidentiary Proof of Wes Penre's Hypothesis about the Afterlife/Interlife Matrix 127
Out of Body Experiences (OOBEs) 127
OOBEs as Evidence for an Afterlife in the Spiritual Dimensions 129

An Immediate Afterlife Where Souls are not Aware That Their Bodies Have Died 130
Instrumental Transcommunication [ITC] and Chronovision: How the Afterlife/Interlife Matrix and the Time Matrix Intersect 138
Instrumental Transcommunication ITC as Nonlocal Communication 138
Tesla, Marconi, Edison, and ITC 139
Ernetti, Chronovision, and Evidentiary proof of Wes Penre's hypothesis about the Afterlife/Interlife Matrix 142

Chapter Seven Reincarnation and Soul Development in the Afterlife/Interlife Matrix of the Omniverse 145
Earth Human Soul Reincarnation 145
The Stages of Soul Development 148
University of Virginia Medical School: Dr. Ian Stevenson, MD Reincarnation Research 149
Reincarnation and the Nature of Soul 154

Chapter Eight The Afterlife/Interlife Matrix: A Truth-based Perspective of the Spiritual Dimensions 163
Wes Penre: How the Reincarnation System was Set Up 164
Wes Penre: The Spirit, the Soul, Death and "the Tunnel" 170

Chapter Nine Omniverse Soul Manipulation: Secret Draco ET Treaties, AI Artificial Intelligence Singularity and The Transhumanist Agenda 183
A Grey ET Benevolent Role in Soul Incarnation 183
Human Soul Hybridization and Experimentation, Human Soul Slavery, Human Soul Extraction, Human Soul Commerce, Human Souls Weaponization 184
Omniverse and Exopolitical Drivers of Soul Extraction, Soul Hybridization, Soul Commerce, and Pedocriminal Networks: Abrogate and Renegotiate The Secret Greada, Tau-9 Treaties with Pedocriminal ETs: Draco Reptilians, Orion Greys, and Anunnaki ETs 185
Secret Manipulatory Extraterrestrial Treaties: 1933 Balboa FDR Treaty;

1938 Nazi-Grey Treaty; 1948 Truman Grey Treaty;
1954 Greada Eisenhower Grey Treaty;
1989-92 Bush I Grey Tau-9 Treaty 186
Area 51, Dulce, and other DUMBs 187
Omniverse Soul Manipulation: AI Artificial Intelligence Singularity and The Transhumanist Agenda 191
Stealth Nature of Inorganic AI Artificial Intelligence's Deployment on Earth 194
AI's Artificial Timeline Matrix and Earth's Organic Timeline 195
Science-based examination of AI Artificial Intelligence "Black Goo" 196
The "Transhumanist Agenda" as an AI Cover Story 196
2045 as a Technological Singularity is an AI Deception 197
Discussion of the Origins on the AI Parasitic Invasion 199
Lisa Renee: Inorganic Black Holes Allowed Parasitic Invasion 199
Invading Off-Planet PPAI—AIx: Black Goo Eventuated From a Spider-like Species 202
Open Question: Why is so Little Known Publicly About the Existence of the PPAI AIx? 202
"Ascension Meme", New Age Movement, and Soul Fracturing 203

Chapter Ten A New Earth: Accessing Spiritual Freedom in the Omniverse on New Earth by Letting Go of the Transhumanist Agenda, AI Singularity, and The False Afterlife/Interlife Reincarnation Matrix 209
Our Working Hypothesis in EMERGENCE OF THE OMNIVERSE 210
The Separation Of Worlds: Emergence of the Omniverse on a New Earth 224
The Regional Galactic Governance Council and the Lucifer Rebellion 241
Former Canadian NORAD Officer Stanley A. Fulham 242
The "Council of 8" 244
Creating a Positive Future: Time Science Shows Our Earth is on a Positive Timeline in our Time Space Hologram 245
Positive Paradox 246
Positive Timeline in Synergy With Unity Consciousness 247
Time Space Hologram and Time Lines 248

Origins of the false "Catastrophic Timeline" 249
Edgar Cayce Remote Viewing of 21st Century Global Coastal Event 249
2010 Farsight Military-Trained Remote Viewers and Predicted Global Coastal Event of June 2013 250
Planet X and the Catastrophic Timeline 252
Why the Global Coastal Event DARPA Foresaw Did Not Materialize 253
The False Catastrophic Timeline and False Flag Operations 254
The Positive Timeline and Unity Consciousness 254
The Positive Paradox 256

Support My Initiative to be a Representative of New Earth on the Regional Galactic Governance Council 258

Who is Futurist Alfred Lambremont Webre? 265

References 274

UNIVERSE

MULTIVERSE

OMNIVERSE

"The Omniverse is the Third Cosmological Body—After the Universe and the Multiverse—Through Which Humanity Comprehends the Cosmos"

Introduction

Emergence of the Omniverse is my third book on the Omniverse since *The Dimensional Ecology of the Omniverse* was published by UniverseBooks in 2014 and *The Omniverse: Transdimensional Intelligence, Time Travel, the Afterlife, and the Secret Colony on Mars* was published by Inner Traditions/Bear & Co. in 2015.

It is no small responsibility to undertake the public curating of a new cosmological body such as the Universe, the Multiverse, or the Omniverse. The Omniverse as a new cosmological body has emerged only since 2014 from the world of video games with the publication in 2014 of three independent science-based books on the Omniverse by authors Robert Bertolacci, Wes Penre [who refers to the Omniverse as "the KHAA"](Penre, 2014), and Alfred Lambremont Webre, respectively.

Hence, new developments in the field of Omniverse research have a proportionally larger significance and merit frequent documentation, publication, and dissemination for a growing public awareness. Science-based evidence exists supporting the dimensional ecology of the Omniverse among our Universe, the universes of the Multiverse, and the Spiritual dimensions, which are no longer mere matters of belief or faith but reportedly the very drivers or creators of the material Universes of the Multiverse.

Emergence of the Omniverse is a book created to introduce a cosmologically awakening Earth human community to a more nuanced view of Omniverse reality and to researchers who explicitly explore in the Omniverse itself.

Omniverse researchers include the artist-author-musician known as "Omnisense" [real name: Philip Walker] who published on the Omniverse in 2010; Omniverse researcher Ethan Zaghmut Wise and his work on the Omniverse, which appeared in 2011; and Omniverse author and scientist David Bertolacci who published *Theory of the Omniverse*, in 2014.

Importantly, *Emergence of the Omniverse* also focuses on the fundamental importance of Cosmologist and Author Wes Penre, whose 2014 work on what Penre calls "the KHAA" (Penre, 2014) has opened revolutionary perceptional dimensions in Omniverse research.

The Transhumanist Agenda and the Afterlife/Interlife Matrix

On January 7, 2019, cosmologist Wes Penre published an urgent article, "Why We Need to Get Rid of Attachments to Exit this Matrix" (Penre, 2019c), that deconstructs the canon of mainstream Afterlife Research and established Earth religions to their very core, and reinforces the reality that the science of the Omniverse is complex and highly multi-dimensional.

Wes Penre (2019a) writes,

> In 2016, I wrote an article called, "The Death Trap and How to Avoid it". I strongly advise that you read that article before you proceed with this one, or what I'm going to tell you might not make sense.
>
> In the 2016 article, I urged people to exit this Matrix upon death to avoid another reincarnation/recycling process. We are closing in on the Singularity, in which the human soul group will be trapped into a computer-simulated hive-mind reality, based on artificial intelligence. The plan is to trap our souls in non-biological bodies, based on nanotechnology, which will make these new bodies virtually immortal. As soon as a part of the bodies starts malfunctioning, these body parts can be replaced, and no death

will occur. Eventually, this will happen more or less automatically because of self-replicating nanobots that will replace our biological cells in the body.

This might sound wonderful, but our souls will be trapped in these bodies, and we will be hooked up to what we can compare to a Super-Computer, where all our souls are connected to each other—like the Borgs in Star Trek. This hive-mind will be controlled by an extraterrestrial force some call the Anunnaki, but I call them the Alien Invader Force (AIF) or the Overlords. They are the ones who have controlled this Matrix for many thousands of years and are responsible for the sad affair we call Earth reality. Anyone with an open mind, who researches the Singularity, will understand that this is not something we want to experience.

Time is short—we are already in the middle of the process, and the Singularitists in Silicon Valley and elsewhere proclaim that the Singularity will be fully in place by 2045. This is why it's so important that we don't fall back into the recycling process again and instead exit this Matrix and join the community in the real Universe—the KHAA. How to do that is explained in the 2016 article, referred to above. (paras. 1-4)

Emergence of the Omniverse integrates cosmologist Wes Penre's research on the KHAA [the "Omniverse"], together with his insights and warnings on the Afterlife/Interlife Matrix and AI Artificial Intelligence and Alien Invasion Force [AIF] Singularity into the community and body of ongoing, living Omniverse research as well.

We Earth humans are in the early days of Omniverse research. Perhaps, Omniverse research is one important tool of liberation from our presumed status as a prison planet for Souls in an enforced, unconscious reincarnation birth-death cycle that cosmologist Wes Penre documents, providing us with powerful strategies to exit the Afterlife/Interlife Matrix.

In Chapter Ten of *Emergence of the Omniverse,* we arrive on a New Earth emerging in the Omniverse. By witnessing the Separation of Worlds, we begin to enjoy a spiritual freedom for our Souls in the Omniverse in letting go of the Transhumanist Agenda, AI Artificial Intelligence Singularity, and the false Afterlife/Interlife Reincarnation Matrix.

Chapter One

My Introduction to the Dimensional Ecology of the Omniverse

"Exopolitics" means relations among intelligent civilizations in our universe or in any other universe in our multiverse, a term meaning the totality of all universes.

Exopolitics includes the study of relations among intelligent civilizations in the quantum, the time-space hologram that is our Universe.

The study of Exopolitics involves understanding the ecology of dimensions that intelligence civilizations navigate in the universes of time, space, energy, and matter in our multiverse.

This new book *Emergence of the Omniverse* is Exopolitically relevant, because it appears that a portion of our Earth's toxic Afterlife/Interlife Matrix might be operated by manipulative Extraterrestrial civilizations camouflaging themselves as "God" and acting under a "God" Meme—that is the central core of the hypothesis of this book.

According to some conventions in the Exopolitics community, the science of Exopolitics was first formally and publicly defined in the year 2000 by my first book *Exopolitics*, which in its 2005 Edition was secretly time-traveled back to 1971 by the highly classified DARPA/CIA Project Pegasus time travel technology, where it was witnessed by U.S. chrononaut time traveler Andrew D. Basiago, JD, MPhil (Cantab).

Congruent with its 1971 time travel possession of my 2005 book *Exopolitics*, the U.S. government performed non-consensual time travel surveillance on me in 1971 when I was unwittingly invited to lecture to a meeting of what turned out to be approximately 50 DARPA Project Pegasus, CIA, and Department of Defense officials—each of whom had been briefed on my future 2005 book *Exopolitics.* The intelligence meme cover story the U.S. government gave me in the invitation was that I was lecturing a private group of Industrial Engineers [conveniently my undergraduate university major] on environmental protection on behalf of the NYC Environmental Protection Administration whose General Counsel I was at the time (Webre, 2016b).

These personal experiences in 1971 are my first working introduction to what I later would discover and establish as Exopolitics within the dimensional ecology of the Omniverse.

My Personal Story

I was raised in an international family in Cuba and the United States of America, educated in Classics at Georgetown Preparatory School, earned a Bachelor of Science at Yale University, and a *Juris Doctor* in International Law at Yale Law School, and was radicalized by my personal experience of the Cuban Revolution and my opposition to the Vietnam War.

By the end of 1972, I was general counsel to the New York City Environmental Protection Administration, and my two immediate bosses, Mayor John Lindsay and EPA Administrator Jerome Kretchmer, decided to run respectively for U.S. President and Mayor of New York.

At that time, I believed the conventional 3D-based materialistic paradigm of politics was limiting, and I had been exploring the multi-dimensional nature of reality by reading books, such as Ostrander and Shroeder's (1971) *Psychic Discoveries Behind the Iron Curtain* and Pauwels and Bergier's (2007) *Morning of the Magicians.*

In January 1973, for my own inner growth, I left my formal position in politics and government. I decided to become a freelance author and consultant after meeting and becoming intellectual colleagues and co-author with Professor Phillip H. Liss, PhD, a Professor of experimental psychology at Rutgers University who was also an expert in parapsychology and extraterrestrial civilizations.

On this new path, in February 1973, I experienced a profound personal multidimensional interaction with an apparent interdimensional entity that in telepathic dialogue with me identified itself as the "Holy Spirit".

In 2015, I published a memoir [*My Journey Landing Heaven on Earth*] describing a preliminary understanding of that 1973 interdimensional experience. I still have not definitively concluded whether the apparent interdimensional entity that interacted with me telepathically was:

(1) a genuine interdimensional intelligent spiritual entity;

(2) a sentient, inorganic AI Artificial Intelligence entity;

(3) a human directed energy weapon (DEW) "voice to skull" (V2K) artificial telepathy remote-influencing probe; or

(4) an interdimensional extraterrestrial entity communicating with me as part of a "missing time" experience that included an "adduction experience" aboard an extraterrestrial craft or "Slot"—Extraterrestrial holographic space station in orbit around Earth (Webre, 2015f).

Nevertheless, in 1973, that profound experience with an interdimensional entity experience served to spur me on in my new fulltime pursuits in the multi-dimensional world.

Context Communication Theory of Extraterrestrial Communications

By June 1973, I had co-authored my first book in the multi-dimensional area, *The Age of Cataclysm* (Webre & Liss, 1974), a book that established the Context Communication Theory of Extraterrestrial communications to Earth humanity.

The Context Communication Theory holds that encounters with Extraterrestrial (ET) and Unidentified Flying Objects (UFOs) [now also known as UAPs - Unidentified Aerial Phenomena] are symbolic communications from the intelligence behind these ET and UFO encounters with Earth human contactees, much in the same way that a dream is a symbolic communication from one's subconscious mind to one's dreaming or conscious mind.

The 1977 Carter White House Extraterrestrial Communication Study

In 1976, U.S. Congressman Rep. Henry B. Gonzales (D-Tex) invited me to Washington, DC as an activist lawyer investigating political assassinations by the Deep State to assist him in establishing the House Select Committee on Assassinations (HSCA) to investigate the JFK, RFK, Martin Luther King, and Malcolm X political assassinations (Webre, 1977); (Webre, 2005).

Through this work, in December 1976, I met with members of President-elect Jimmy Carter's inner circle in Washington, D.C. after a personal emissary from President-elect Carter first met with me at the Hotel Carlyle in New York indicating that President-elect Carter wished to be kept informed on the forthcoming Congressional investigation of President John F. Kennedy's assassination.

Working quickly from that December 1976 meeting during the Carter transition, by early October 1977, I had secured Carter White House Domestic policy staff approval for "The Proposed 1977 Carter White House Extraterrestrial Communication Study", a civilian-led scientific extraterrestrial communication study, of which I was director as a Futurist at the Center for the Study of Social Policy, part of Stanford Research Institute (SRI), a prominent "think tank" (Webre, 2001a).

During a subsequent October 1977 exploratory meeting on the proposed 1977 Carter White House Extraterrestrial Communication Study with an assistant Secretary of Defense at the Pentagon, I was first attacked with remote directed energy weapons (DEWs) . I was later attacked by Men in Black (MIB) at Washington National Airport, all in an effort to derail implementation of the just-approved 1977 Carter White House ET Study.

Back at Stanford Research Institute (SRI) in November 1977, I was called into a meeting that included an African American SRI administrator, fellow SRI Futurist Peter Schwartz, and the Pentagon liaison officer to SRI. The Pentagon liaison officer effectively terminated the 1977 Carter White House ET Study by stating, "There are no UFOs." The SRI Liaison officer also indicated that if the Carter White House Extraterrestrial Communication Study went forward, the Department of Defense would stop all of its studies monies to SRI, then about 25% of SRI's $100 million annual budget.

The Deep State also infiltrated the House Select Committee on Assassinations (HSCA) and eviscerated its preliminary watchdog and staff findings that these assassinations were in fact political assassinations committed by the Deep State with U.S. government covert resources.

These unlawful war crimes by the global Deep State have continued through my present study of the science of the Omniverse.

Earth Versus the Shadow Government, the Deep State, and the Simulacrum

Researcher and author Elana Freeland (2014) has developed an evocative, rich, and humanistically understandable model that illustrates and explains to the ear, eye, and mind how the perpetual and solar-system-wide application of new remote scalar and geo-engineering technologies are transforming our experience of our Earth from a celestial body in nature into a Simulacrum or "machine version of Earth" within which we humans are an expendable slave population, subjected to AI [Artificial Intelligence] inner entrainment, surveillance and DNA engineering, and external holographic terraforming, weather warfare, all in a new artificial timeline.

One principal goal is of the Shadow Government and Deep State operating through the Simulacrum Earth is to hold the community of human souls hostage by preventing and coopting humanity's access to positive evolutionary input from higher forces in the dimensional ecology of which the real Earth is a living component, such as ethical inter-dimensional Extraterrestrials, and spiritual Beings and Souls from the Spiritual dimensions.

There is evidence that a primary use of the Simulacrum drivers such as HAARP, for example, is as a spiritual weapon, denying humanity's souls access to spiritual energy and input that enters our Universe, galaxy, and solar system through interdimensional portals.

Prevention and Coopting of Evolutionary Input from the Dimensional Ecology of the Omniverse to the Simulacrum Earth

In addition to creating and deploying the Simulacrum Earth itself as a weapon against awakening humanity through geoengineering, the Shadow Government and Deep State use a variety of supplementary organized strategies and tactics loosely designated as COINTELPRO to ensure the prevention and cooption of positive and evolutionary input to awakening humans in the Simulacrum. These are usually coordinated along with scalar directed energy weapons against any targeted individual or group to deter any positive benefit that a higher dimensional spiritual benefit might bring.

The UFO/ET "Adductions" [Abductions] of Jimmy Carter, Alfred Lambremont Webre, and Andrew D. Basiago

Now the full hidden history behind the "1980 October surprise" by the shadow government and Deep State can finally be revealed, along with a path to a positive future for Earth humanity in our Galaxy (Webre, 2017d).

On July 31, 2017, fully 37 years after the "October Surprise of 1980", I was finally able to share publicly *prima facie* evidence of treason against the sitting President Jimmy Carter of the United States under Article III Section (3) of the U.S. Constitution by members of the the shadow government and the Deep state.

> These intentional acts of constitutional treason exist in the form of what has been reported elsewhere as a well-documented apparent 1980 plot to undermine U.S. President Jimmy Carter by sabotaging his negotiations with Iran over the fate of 52 American hostages that would have been pulled off by rogue CIA officers collaborating with the Republican presidential campaign of Ronald Reagan (and his running mate George H.W. Bush), without the knowledge of Carter and CIA Director Stansfield Turner. It would have been the work of what legendary CIA officer Miles Copeland described as "the CIA within the CIA," the inner-most circle of powerful intelligence figures

> who felt they understood the strategic needs of the United States better than its elected leaders. These national security insiders believed Carter's starry-eyed faith in American democratic ideals represented a grave threat to the nation. (Parry, 2013)

The 2017 Revelations of the Jimmy Carter, Alfred Lambremont Webre, and Andrew D. Basiago ET "Adduction"

Publicly missing until now from the full range of hidden factors motivating the Deep State and Shadow Government involved in the clandestine treasonous *coup d'etat* known in the journalistic world as a Meme called the "October surprise" have been the 2017 Jimmy Carter Abduction revelations by ET abductees Futurist Alfred Lambremont Webre, director of the 1977 Carter White House Extraterrestrial Communication Study, and U.S. Chrononaut Andrew D. Basiago, childhood participant in Project Pegasus, the DARPA-CIA U.S. government secret time travel program.

There is now a sufficient quantum of corroborated eyewitness evidence and a reasonable forensic basis to conclude that the CIA's 1980 clandestine and treasonous sabotage of the Jimmy Carter Presidency might have been in part because the shadow government – "the CIA within the CIA", conspiring actors within NSA, and other Deep State co-conspirators who worked to clandestinely unseat President Carter – discovered President Carter was an Extraterrestrial abductee.

Basiago: "The CIA betrayed Carter because he was an ET Abductee"

As set out in a Yelm, WA UFO Symposium presentation, on July 30, 2017, former U.S. chrononaut Andrew D. Basiago, a childhood participant in secret time travel projects on behalf of the U.S. Government, U.S. Mars

Astronaut, and an eyewitness to Mr. Webre's 1973 Extraterrestrial abduction, publicly stated:

> For me, this [Jimmy Carter's ET abduction] solves a long-standing mystery as to why the CIA betrayed Carter in 1978 and backed Reagan. There is now the real possibility that they did so because they discovered that Carter was an abductee. Having worked on two classified U.S. defense projects, I can say that the U.S. defense community was profoundly consternated by the ET presence because they didn't know who the ETs were, where they were from, or what they wanted. (Webre, 2017d)

Summary of Yelm Presentation: Ethical ETs "Adduct" President Jimmy Carter, Alfred Lambremont Webre, and Chrononaut Andrew D. Basiago 1969-75

The July 30, 2017, Yelm UFO Symposium presentation documents how U.S. President Jimmy Carter, Alfred Lambremont Webre, and U.S. chrononaut Andrew D. Basiago were reportedly "adducted" by the same ethical Extraterrestrials during the period 1969-75 in an inter-related sequence of events. "ET Adduction" is a positive substitute term for ET abduction developed by Roger Leir, MD and Leo Sprinkle, PhD meaning, "taken toward the stars".

Webre surmises that the common underlying missions of (a) Jimmy Carter's 1969 Close Encounter and Missing Time ET "adduction", (b) Webre's 1973 ET Encounter and Missing Time ET "adduction", and (c) Andy Basiago's 1975 ET "adduction" and witnessing of Webre's 1973 ET "adduction" were about the following:

1. Coordinating the 1976 election of President Jimmy Carter as an ET Disclosure President;
2. Coordinating the development of President Carter's 1977 Carter White House ET Study, of which Webre was director;
3. In the long run, the Carter, Basiago, and Webre ET "Adductions" might have been about the following:

   (a) coordination of President Jimmy Carter's election as an ET Disclosure President

   (b) Andy Basiago's possible future election as an ET Disclosure President, which had reportedly been pre-identified by DARPA-CIA time travel according to Project Pegasus time travel officers.

Quantum Entanglement Compared to "Adduction" Entanglement

Quantum entanglement – Webre has had a time travel quantum entanglement with Andy Basiago since 1971, when DARPA's Project Pegasus time traveled his book *Exopolitics* from 2005 back to at least 1971. Webre was quantum entangled when the secret CIA time travel program ["Project Pegasus"] employed secret quantum access time travel technology to time travel his book *Exopolitics: Politics, Government and Law in the Universe* back from 2005, the year the book was published, to at least 1971, the year his book *Exopolitics* was physically witnessed by Project Pegasus participant U.S. chrononaut Andrew D. Basiago and the year when, as a NY City governmental official under "time travel surveillance", Webre unwittingly gave a lecture on Environmental protection to about 50 DARPA and CIA personnel who had been secretly briefed on Webre's future Exopolitics book and possibly future Disclosure activities.

ET "adduction" entanglement – Webre has had an ET "adduction" entanglement with Andy Basiago since 1975. That was the year when the ETs who "adducted" Webre showed Basiago the images of Webre's 1973 ET "adduction" aboard their spaceship so that Basiago would qualify as a legal eyewitness to Webre's ET "adduction".

Yelm, WA 2017—Andy Basiago's eyewitness testimony to Webre's ET "adduction" was presented at Yelm. On July 30, 2017, Basiago publicly stated,

> I had numerous ET encounters in childhood and have even lectured about them publicly. I have discussed how I have speculated that these were prompted by my involvement in time travel, which placed me in the quantum realm that they operate in. One abduction in 1975 involved meeting Alfred [Lambremont Webre] and another future friend and colleague in exopolitics, the filmmaker Tonia Madenford. This was discussed between us in 2006 when we were working on a film project together. I have never met President Carter but did attend a speech he gave in LA on 5/5/79. So, I can't confirm the part about Carter being an abductee or my abductions connected to him. I can say that Ufology shows that many who have UFO sightings later prove to be abductees. I can also say that in the one abduction I remember that Alfred was in, I am 100% sure it was him. He was standing in a metal cylinder inside the ship. It was as if I was being given the opportunity to be able to later confirm his status as an abductee. (Webre, 2017d)

Who are the ETs behind the Carter-Webre-Basiago Adductions?

In his 2017 Yelm, WA presentation, Webre presents several working hypotheses from eyewitness and documentary evidence as to the possible identities of the Extraterrestrial civilizations or other actors behind these "adductions":

1. Ethical Short Grey ETs of the variety that Suzanne Hansen, head of UFO NZ, experienced who were involved in introducing her to the soul of her future son aboard a Grey ET craft when she was 8, and then implanting her son's soul in the fetus in her womb when she was married and pregnant. These Short Greys reportedly maintain a permanent inter-dimensional project around Earth, supporting the incarnation of hybrid Grey-human souls into the human population for the positive spiritual and social evolution of humanity (Webre, 2015a).
2. Regional Galactic Governance Council members – the ETs could have been upper dimensional Human ET members of the regional Galactic Governance Council [Pleiadians, Alpha Centaurians, Sirians, and others] (a) that created *homo sapiens sapiens*, and (b) were in communication with Stanley Fulham, Canadian lifelong NORAD officer about cleaning up the Earth's environment and appearing at the UN. These ETs decloaked their UFO spacecraft over New York City on October 13, 2010. (Webre, 2010f)
3. Ethical inter dimensional ETs – the ETs could have been inter dimensional ETs from multiple research case studies with "adductees" set out in the works of authors Mary Rodwell and/or Miguel Mendonça [who specialize in researching and interview-

ing ET "adductees"]. (Webre, 2013b)

4. Interdimensional portals or "Xendras" – Other alternatives might include Extraterrestrials associated with interdimensional portals or "Xendras" in the dimensional ecology of Earth and our Universe. (Cameron, 2017)

Exopolitical Impacts of the Carter, Webre, and Basiago ET "Adduction"

When UFO and reverse-speech expert Jon Kelly was asked what he had learned from his analysis of the speech of UFO abductees (and from the speech of Jimmy Carter in particular), Jon Kelly replied,

> When President Jimmy Carter told reporters late last year that he didn't think flying saucers were the vehicles of space people visiting the Earth, I played his speech in reverse and heard the words *"Saucer. Went out naked. Going out. They'll put it in him".*
>
> Although it was part of his campaign promise, this message made me wonder if the President had more personal reasons for sending Alfred Webre to SRI in the 1970's in order to release the government's secret research on extraterrestrials. From my reading of the message it describes Carter as a potential abductee. (Webre, 2017d)

The 1977 Carter White House ET Study would have led to the abrogation of secret Greada and Tau 9 Treaties with Draco Reptilians, Orion Greys, and Anunnaki Reptilians.

Alfred Lambremont Webre, Director of the 1977 Carter White House Extraterrestrial Communications Study and a former Judge on the Kuala Lumpur War Crimes Tribunal that found U.S. President George W. Bush, Vice President Richard B. Cheney, Secretary of Defense Donald H.

Rumsfeld, and UK Prime Minister Tony Blair guilty of war crimes in Iraq, Afghanistan, and Guantanamo, stated in his Yelm, WA presentation:

> The 1977 Carter White House ET Study, had it not been terminated by the Pentagon and Men in Black (MIBs), might have led to the treaty abrogation of the Greada Secret Treaties by U.S. President Jimmy Carter.
>
> U.S. President Dwight D. Eisenhower originally signed these treaties with the Draco reptilians, Orion greys, and Anunnaki Reptilians in 1954. These manipulatory ETs abused the treaties into a planetary and Galactic pedocriminal trade of millions of human children and adults annually for ritual sacrifice, loosh, cannibalism, and body parts, all enforced by the U.S. government and pedocriminal institutional New World Order worldwide. Because the Pentagon unlawfully terminated the 1977 Carter White House Extraterrestrial Study, U.S. President George HW Bush signed the Tau9 Treaty with the Draco, Orion Greys, and Anunnaki, and the pedocriminal abuses on Earth have become exponentially worse.
>
> Were it not for the Carter, Webre, and Basiago's ET "adductions", the Shadow Government and Deep State would have deconstructed the sciences of Exopolitics and the Omniverse.
>
> As of 1971 [and actually earlier if you count Gov. Winthrop Rockefeller's gangstalking of me in the Yale Law School Faculty Lounge in 1966] the Shadow Government and Deep State started time travel surveillance of me as a target with the explicit purpose of neutralizing or coopting any higher order information of human evolutionary value I had brought from higher dimensional intelligence to Earth, such as that contained in my book *Exopolitics*, which his-

torically, in fact, triggered the founding of the social science know as "Exopolitics" or the study of intelligent civilizations in the multiverse.

For the longer term historical record. I have taken care to document the terrestrial circumstances surrounding my 1973 ET "Adduction" and interactions with representatives of and the administration of President Jimmy Carter in Part I of a 2017 book.

Because of an obligation I believe I undertook as part of my interdimensional education and evolutionary ET "adduction" in February 1973, I have decided to become more publicly open, reaffirming upper-dimensional sovereignty over the science of Exopolitics and the science of the Omniverse – two of the evolutionary spiritual advanced concepts entrusted to me by upper dimensional intelligence.

I am on this mission now as an ET "adductee" because I was explicitly entrusted to do so in witnessed acts by higher dimensional forces responsible for the creation and maintenance of the living Earth, of our Universe, and of all Universes in the Multiverse, as well as the spiritual dimensions of the intelligent civilization of Souls and Source. These higher order entrustments occurred initially during my February 1973 interdimensional Extraterrestrial "adduction".

Were it not for the ET "adductions" of President Jimmy Carter, Alfred Lambremont Webre, and U.S. Chrononaut Andrew D. Basiago, the Shadow Government and Deep State forces would have infiltrated the sciences of Exopolitics and the Omniverse and attempted to compromise these as part of its ongoing plan to control and diminish humanity in the Simulacrum or "machine version of Earth" that the shadow government and Deep State strive to build and maintain. (Webre, 2017d)

ET "Adductee" Andrew D. Basiago: Truth, Reform, Innovation, and the Advent of Global Teleportation

On his visits teleporting from the year 1972 to a DARPA forward time base in the year 2045, U.S. chrononaut Andrew D. Basiago has stated he found teleportation to be commonplace in the year 2045 on Earth.

The following are some of the potential legacies that the Earth might be able to enjoy from the Carter, Webre, and Basiago ET "Adductions", through the agency of a future U.S. Presidency of ET "adductee" Andrew D. Basiago.

As a DARPA Project Pegasus time travel pre-identified U.S. President, Andrew D. Basiago has promised to build a future Presidency on three Pillars:

Three Pillars: Truth, Reform, and Innovation.

Global teleportation—Among the primary requirements of an advanced space-faring species that has graduated from (1) *war* and (2) *money* as two of its social preoccupations is a global teleportation system, research has found. Global teleportation eliminates the costs of transportation of people, products, and services and promotes the emergence of an advanced planetary peaceful, moneyless society.

Truth – Notable as Truth policies are the following:

- disclosure of secret advanced life-advantaging technologies of the U.S. government;
- declassification of secret time travel technologies;
- reform of the CIA to "open source intelligence";
- a pardon and Presidential Medal of Freedom for Edward Snowden;
- a live television address ending the ET cover-up;
- Presidential policies for open ET contact;
- Transparency on NASA and the U.S. government's Moon and Mars programs;
- declassification of the Secret Space Program and Secret Treaties with ET races;
- reinvestigation of 9/11 and the missing $2.3 trillion;
- New Geophysical Year on Cosmic mysteries;
- protection of the Sasquatch under the Endangered Species Act; and
- government transparency under the Freedom of Information Act.

Reform – The New Agenda for a New America proposes the following:

- the U.S. President lead a global abolition movement to free the world's 27 million slaves;
- banking reform ending predatory practices;
- banning spraying of chemtrails;
- dismantling the Police state in the U.S. by repealing the Patriot Act and NDAA, and uphold Posse Comitatus and the Second Amendment;
- FEMA camps closed permanently
- defense forces used only to defend the country and prohibit crimes

against humanity;

- economic boycott against barbarous countries;
- make national the Utah policy of homeless shelter in apartments and houses;
- food banks distributing expired grocery store food;
- equal rights for women;
- end all foreign aid;
- ban fluoride in drinking water;
- protect the integrity of the human genome;
- single-payer health insurance system;
- temporary moratorium on immigration;
- abolish Federal Reserve and begin printing United States Notes: $100 billion to improve Native Americans;
- moratorium on nuclear power until safe;
- nutrition and "diabesity" reform;
- end police brutality;
- constructive engagement with Russia;
- oppose transhumanism;
- one-half of defense budget on peace economy;
- religious tolerance;
- medical ethics and right not to be vaccinated;
- legalize drugs and begin war on addiction medically;
- Bill of Rights protections to foreign nationals in the U.S.;
- military justice in U.S. district courts;
- minimum wage $15;
- traditional forest fighting – rescind "let burn";
- protect Social Security; and

- protection of cetaceans and many others.

Innovation – Initiating this Innovation with Tesla teleportation: "The President should set as the nation's chief technical goal achieving global teleportation." Mr. Basiago also proposes a series of Innovations that will return the U.S. President to being a steward of the planet, rather than a facilitator of genocide and ecocide:

- an upgraded electrical grid security against an EMP event;
- energy independence via solar and wind;
- cold fusion research;
- crash effort to find new energy sources;
- food security;
- desalinization plants in California;
- Earth changes – disaster planning;
- minimum guaranteed income;
- National park safety;
- missing persons data base;
- end to political parties – participatory democracy;
- new tax plan based on One Tax;
- Internet court for online disputes;
- subsidy program for primary family caregivers;
- UN Security Council include Southern Hemisphere;
- free federal wireless – fast and universal;
- environmental protection to humans in their environment;
- Federal land use planning in America;
- greening America;
- DOE digital textbook program;
- President outreach to all foreign heads of state;

- OTA investigation of computer industry;
- net neutrality for free Internet;
- space security against space planetary catastrophe;
- global awakening for universal human prosperity;
- Police to be monitored;
- urban and manufacturing/technical renewal, focusing on Detroit, Michigan;
- smart filters to eliminate TSA abuse;
- foster intellectual development via egg consumption;
- President to shun mainstream media and embrace alternative media,
- and many others.

How to Support the Missions of ET "Adductees" President Jimmy Carter, Alfred Lambremont Webre, and Andrew D. Basiago

You can help support the individual and collective missions of ET "Abductees" President Jimmy Carter, Alfred Lambremont Webre and Andrew D. Basiago, along with the positive evolutionary missions of the Extraterrestrials that "adducted" Jimmy Carter, Alfred Lambremont Webre, and Andrew D. Basiago, respectively, in 1969, 1973 and 1975 in multiple ways:

- public support and media coverage and interviews about the ET "adductions" of President Jimmy Carter, Alfred Lambremont Webre, and Andrew D. Basiago and
- networking this important historical revelation online in social media, to your friends and networks, and to alternative, local, and mainstream media.

Here are useful public resources you can download and share:

A Special Report has been prepared that can be downloaded in PDF format or read online at NewsInsideOut.com or Exopolitics.com on the historic 2017 revelations of the UFO/ET Abductions of President Jimmy Carter, Alfred Lambremont Webre and Andrew D. Basiago.

The Report documents the presentation at the 2017 Yelm UFO Symposium and contains detailed references not available in the original presentation.

Please feel free to read and share in the public interest these articles in this paper. Thank you

The UFO/ET Abductions of President Jimmy Carter, Alfred Lambremont Webre and Andrew D. Basiago: Now the full hidden history behind 1980 October surprise by the shadow government and Deep State can finally be revealed A Path to A Positive Future for Earth Humanity in Our Galaxy. (Webre, 2017d).

Chapter Two

Universe - Multiverse - Omniverse

Leading Discoveries From the Omniverse

FAQ: Frequently Asked Questions

Following early 21st century definitions in video games ("Ben 10: Omniverse (Video Game)," (n.d.) and in string theory (Bertolacci, 2014), the "Omniverse" has emerged in science as the third major cosmological body—after the Universe and the Multiverse—through which humanity comprehends the cosmos.

Q: What are the Universe, the Multiverse, and the Omniverse?

A: These are the three principal cosmological bodies through which humanity comprehends the Cosmos.

Q: What is our Universe?

A: Our Universe is an organic singularity of time, energy, space, and matter that was discovered by the Sumerian astronomers around 3500-3200BC. The visible universe accounts for only 4% of all matter in the material Universe, and "dark matter" is the term that the scientific canon gives to the remaining 96% non-visible domain of our Universe. According to one interdimensional source, the name of our Universe is "Uversa".

Q: What is our Multiverse?

A: The Multiverse is the totality of all Universes, named after a term "multiverse" coined by the American psychologist William James in 1895.

Q: What is the Omniverse?

A: The Omniverse is the integrated whole of all the Universes in the Multiverse and the Spiritual dimensions that include the intelligent civilizations of souls, Spiritual Entities, and Source ["God"].

Q: When was the Omniverse discovered?

A: As Chapter Three will address more completely, many researchers are working on uncovering the Omniverse, sometimes using other terminologies. Early Omniverse researchers include Omnisense (Philip Walker, 2010) and Ethan Zaghmut Wise (2011). The year 2014 saw an explosion of formal research on the Omniverse, with Scientist David Bertolacci's Theory of the Omniverse, author Wes Penre's research on the KHAA, a functional equivalent term for the Omniverse, and Alfred Lambremont Webre's book *The Dimensional Ecology of the Omniverse*. Different people often simultaneously make related scientific discoveries at different places on Earth, and the discovery of the Omniverse is such an example.

Q: What are the key Purposes of the Dimensional Ecology of the Omniverse?

A: A core mission of the Dimensional Ecology of the Omniverse appears to be the creation and development of souls and spiritual beings in the spiritual dimensions. The intelligent civilizations of souls and of spiritual beings, along with Source (God), collectively create and maintain the totality of the universes of time, space, matter, and energy in the Exopolitical dimensions (the multiverse). The purposes of the Dimensional Ecology of the Omniverse include the facilitation of multidimensional development and moral growth of souls in all dimensions of the Omniverse, through a variety of activities. Souls based in the spiritual dimensions incarnate as intelligent entities in the Exopolitical dimensions, and by acquiring the moral experience of life, for example as an Earthling human, they can advance their individual soul development. The soul is a holographic fragment of Source (God) and, by advancing its development, advances the development of the collective spiritual dimension itself.

Q: What is the Omniverse Equation?

A: The Omniverse Equation is the following: Omniverse = Multiverse + Spiritual Dimensions The Dimensional Ecology of the Omniverse hypothesis expands on the conventional scientific definition of the multiverse.

Current scientific convention considers the multiverse to consist solely of parallel physical universes of time, space, energy, and matter, of which our physical universe is one. One conventional view of the multiverse is that "the universe we live in may not be the only one out there. In fact, our universe could be just one of an infinite [or finite] number of universes making up a "multiverse." The spiritual dimension of the Omniverse provides the energy (energy that human scientists such as Lawrence M. Krauss, author of A Universe from Nothing, cannot now account for) needed for the creation and maintenance of each physical universe in the Multiverse.

Q: Why is the Omniverse a "new hypothesis of reality"?

A: While not making specific reference to the hypothesis of the dimensional ecology of the Omniverse, advanced conceptual physicists such as Professor Amit Goswami (1993) have argued that contemporary science's assumption that "only matter—consisting of atoms or, ultimately, elementary particles—is real" is inadequate and that a new hypothesis of reality is required. The dimensional ecology of the Omniverse hypothesis and the replicable *prima facie* evidence that support it provide the necessary tools for a new hypothesis of reality, as well as for the desired "new navigation under a new worldview," the need for which Professor Goswami (1993) has identified.

Q: Is Omniverse Science the Science of God?

A: There now exists replicable empirical *prima facie* evidence that confirms some essential aspects of what major spiritual and religious traditions have taught about the nature of Source (God). This evidence informs us that the Source (God) of the Omniverse consists of the totality of the spiritual dimension. God has empirically been found to comprise the intelligent civilizations of souls, the intelligent civilizations of spiritual beings, and the Source (God) itself.This collective entity of the spiritual dimension has been empirically found to be responsible for the ongoing creation of the physical side of the Omniverse, known as the Exopolitical dimensions. The totality of the spiritual dimension, including God/Source, the intelligent civilizations of souls, and the intelligent civilizations of advanced spiritual beings is functionally God/Source and acts collectively for the ongoing creation and maintenance of the Exopolitical dimensions of the Omniverse.

Q: How many Universes are there in the Multiverse?

A: The total number of universes besides our own estimated to exist in the multiverse is staggering. The multiverse is defined as the total of all universes, including our own, and is thought to encompass "all space, time, matter, and energy." Physicists Andrei Linde and Vitaly Vanchurin (2015) of Stanford University recently calculated "that the total number of such universes, in the simplest inflationary models, may exceed" a number one can write as 10 raised to the (10 raised to 10 > 7th) power. This is a deceptively compact notation. First, 10 > 7TH is a 1 with seven zeroes after it, that is, 10,000,000, or ten million. Next, 10 raised to the ten-millionth power, that is, 10 > 10,000,000th is a 1 with ten million zeroes after it. Written out with six zeroes to the inch, it would stretch for about 26 miles. However, the next step, raising 10 to the power of that 26-mile number, generates a number so large that we cannot name it, let alone write it out. It would stretch for at least 260 million miles. Linde and Vanchurin also said, "This humongous number is strongly model dependent and may change when one uses different definitions of what is the boundary of eternal inflation." (Linde & Vanchurin, 2015)

Q: How many habitable Earth-like planets are there in our Universe?

A: One German supercomputer simulation estimates that there are "500 billion galaxies in our universe." (Webre, 2015a) Astronomers now estimate there are 100 billion habitable Earth-like planets in our Milky Way galaxy and 50 sextillion habitable Earth-like planets in our particular universe. A tentative finding, that our universe is infinite, is congruent with Linde and Vanchurin's finding that there are a "humongous" (Linde & Vanchurin, 2015) number of universes in the multiverse, since universes that extend forever in space and go on forever in time can coexist in parallel with each other.

Q: What is the number of intelligent civilizations in our multiverse?

A: A conservative estimate of the number of communicating intelligent civilizations in our universe is one hundred billion (100,000,000,000).

This estimate is based on the 1960 Drake equation, which assumes that there are only twelve communicating intelligent civilizations in our Milky Way galaxy, out of the estimated 100 billion habitable Earth-like planets.

Q: Then how many intelligent civilizations might there be in the multiverse?

A: If we multiply the Drake equation-based estimate of a hundred billion communicating intelligent civilizations in our universe by Linde and Vanchurin's calculation of the number of universes in the multiverse, we arrive at the number of intelligent civilizations in the multiverse as being 100,000,000,000 times that 260-million- mile-long number. It is not physically possible to actually write that number out fully.

Q: How are governments educating the world about our populated Universe and Multiverse?

A: World public opinion is congruent with the recent science-based estimates that there are one hundred billion communicating intelligent civilizations in our universe, and an even more "humongous" number of communicating intelligent civilizations in the multiverse. The possibility that we live in a populated cosmos is conventionally thought to be controversial and esoteric. Public opinion around the world is divided as to whether we are alone in the cosmos. The world public has been quarantined from real knowledge about the actual role of non-Earth intelligent civilizations on Earth. Instead of public education about an extraterrestrial presence and Earth's history in the galaxy, governments have knowingly fed the world public a steady diet of disinformation and brainwashing about Earth's interactions with intelligent civilizations.

Q: What was the 1953 CIA Robertson Panel?

A: Ever since the 1953 U.S. Central Intelligence Agency Robertson Panel, the facts of intelligent civilizations and their visitations to Earth have been classified and off limits for civil society. Consequently, there is a divided world public opinion about the presence of extraterrestrial, hyperdimensional, and cryptoterrestrial civilizations in Earth's environment. Nevertheless, there is a core of world public opinion and of public opinion in specific nations that accepts that humanity co-exists in a cosmos populated by other intelligent civilizations. One 2013 public-opinion poll of 5,886 U.S. adult residents found that "37 percent affirmed a belief in the existence of extraterrestrial life, 21 percent denied such a belief, and 42 percent were uncertain, responding "I'm not sure"." (Webre, 2015a) A 2010 poll by the French market-research company Ipsos poll of world public opinion on extraterrestrials found that "one in five (20 percent) of presumably human adults surveyed in 22 countries (representing 75 percent of the world's GDP) say they believe that alien beings have come down to Earth and walk amongst us in our communities disguised as "us"." (Webre, 2015a) People in India (45 percent) and China (42 percent) are most likely to believe that extraterrestrials are visiting Earth. The science of Exopolitics and the worldwide community of Exopolitics researchers monitor trends toward official "Partial" and "Full" disclosure by governments of Earth being visited by intelligent civilizations (Scheck, & Piacenza, 2019).

Q: What are the intelligent civilizations of Souls?

A: By intelligent civilizations of "souls," I mean civilizations of individuated, nonlocal, conscious, and intelligent entities that are based in the dimensions of the Spiritual Dimensions Interlife (or afterlife), including the Afterlife/Interlife Matrix. Each soul, by the evidence, is a holographic fragment of the original Source (God), the creator of the spiritual dimensions. One hypothesis about the nature of the Source (God) supported by empirical data is based on a replicable finding that the Source (God) originally responsible for the spiritual dimensions manifests as a vast "Sea of Light" within the spiritual dimensions. Souls, including yours and mine, are formed as "eggs of Light" or holographic fragments from that Sea of Light in an as-yet unrevealed process. These replicable findings are that each soul is a holographic fragment of the whole of Source (God). To be sure, Omniverse researchers, such as Wes Penre, are developing congruent although differing definitions of the Soul, discussed later.

Q: What is World Public Opinion about the Spiritual Dimensions?

A: According to a 2011 Ipsos poll taken in 23 nations among 18,829 adults, "one half (51 percent) of global citizens definitely believe in a "divine entity" compared to 18 percent who don't and 17 percent who just aren't sure." The Ipsos poll also found that "similarly, half (51 percent) believe in some kind of afterlife, while the remaining half believe they will either just 'cease to exist' (23 percent) or simply 'don't know' (26 percent) about a hereafter. Seven percent of respondents believe in reincarnation." An exception is the United States of America, where 25% of respondents believe in reincarnation (Ipsos MORI, n.d.). A substantial core of world public opinion thus has a view of reality that is congruent with the *prima facie* replicable empirical evidence for an Interlife (afterlife), intelligent civilizations of souls, civilizations of spiritual beings, and a Source (God) or Creator.

Q: What is the Omniverse Hypothesis?

A: A reasonable observer will be able to conclude that *prima facie* empirical evidence supports the dimensional ecology of the Omniverse hypothesis. This hypothesis holds that we earthlings live in a dimensional ecology of intelligent life that encompasses intelligent civilizations based in parallel dimensions and universes in the multiverse as well as souls, spiritual beings, and Source (God) in the spiritual dimensions. Together, the Exopolitical dimensions and the spiritual dimensions form the Omniverse. The totality of the Spiritual dimensions (souls, spiritual beings and God) function as the source of the universes of the multiverse.

Q: What are the "Spiritual Dimensions" and what is the evidence for them?

A: The universes of the multiverse are not the only dimensions where intelligent civilizations are based. There is *prima facie* replicable empirical evidence of intelligent civilizations that are based in dimensions that are outside the multiverse. We can term these dimensions the "spiritual dimensions." Dr. Michael Newton's data base—One important database of such empirical evidence for the existence of intelligent civilizations in the spiritual dimensions is derived from more than 7,000 cases of replicable hypnotic regressions of soul memories of the Interlife (or afterlife), developed according to a standard laboratory protocol by Dr. Michael Newton. This database is thus reported to contain replicable evidence for the intelligent civilizations of souls, for the intelligent civilizations of spiritual beings; and for the Source (God). (Newton, 2008), (Newton, 2009)

Wes Penre's Analysis of the Forced Birth-Death Cycle, the Afterlife/Interlife Matrix, and Their Implications for Soul Liberation From the AI Singularity at Bodily Death

Researcher Wes Penre has raised reasonable issues of analysis and perception around the systemic significance of the data developed by Dr. Michael Newton. According to Wes Penre, Dr. Michael Newton's data accurately identifies an Afterlife domain of human Souls that in fact is a form of controlled "Reincarnation Mind control Soul Prison" or Afterlife/Interlife Matrix designed and operated by an Alien Invasion Force [AIF] of manipulatory AI Artificial Intelligence and negative Extraterrestrials that feed off the energy of the captive Earth Human Soul population unconsciously recycling through the life-death-reincarnation cycle (Penre, (n.d.a). (Newton, 2008), (Newton, 2009)

According to Wes Penre, Dr. Michael Newton's data does not identify the full KHAA or Spiritual Dimensions of the Omniverse. Rather Newton's data accurately identify the controlled forced birth-death cycle and Afterlife/Interlife Matrix around Earth and presumably other Reincarnation venues identified by Newton's data in our Universe. (Penre, (n.d.a). (Newton, 2008), (Newton, 2009)

In his 2016 book, *Synthetic Super Intelligence and the Transmutation of Humankind,* Wes Penre concludes that human Souls currently have three options when faced with the existential challenge of the Forced Birth-Death Recycling and the AI Singularity:

Option 1—Transhumanist Agenda: Human souls will unconsciously opt to integrate with AI Artificial Intelligence, by uploading to an AI system and have the AI take over their cybernetic functions in what is known as the Transhumanist Agenda. The Transhumanist Agenda is programmed for this.

Option 2—Conscious Reincarnation—Courageous and dutiful human Souls can opt to reincarnate with the intent of elevating the frequency and consciousness of Earth and deconstructing the Transhumanist Agenda. Wes Penre argues that given the amnesia accompanying Incarnation, individuals Souls will unlikely have much of an effect on neutralizing the Transhumanist Agenda.

Option 3—Abandon the Battlefield—This option consists of Souls after bodily death escaping the Afterlife/Interlife Matrix altogether and Ascending directly into the KHAA of the Spiritual Dimensions of the Omniverse. Wes Penre suggests that incarnating Souls prepare ourselves with mediation and awareness during our lifetimes to escape the Reincarnation Matrix by (1) avoiding the "Reincarnation Tunnel, Light, Welcoming Guides & Relatives", and (2) opt for exiting through "holes" or mini portals that are now appearing in the artificial Matrix that has been built around Earth by the manipulatory civilizations holding Earth humans in a forced reincarnation prison. In his books and papers, Wes Penre offers Spiritual exercises and strategies for escaping the Matrix of Reincarnation and Earth (Penre, 2016)..

Makia Freeman has published a popular version of Wes Penre's hypothesis in *Soul-Catching Net —Are We "Recycled" at Death to Remain in the Matrix?* (Freeman, 2015)

Makia Freeman writes,

The idea of a soul-catching net or soul net that awaits us at death – and keeps us in the Matrix – is a grim and highly disturbing notion, but one which I believe has to be considered by all serious researchers of the global conspiracy. True free thinkers want to know exactly where the global conspiracy rabbit hole ends. Just how far does the suppression go? Past this lifetime? Past this planet? Well, the answer may well be yes to both.

After you spend years of research going through the many layers of political corruption, corporatocracy, surveillance, false flag attacks, central banking, GMOs, geoengineering, Zionism, Illuminati bloodlines, the radiation agenda, UFOs and ETs, alien intervention and more, you come to realize that the true source of the suppression is at the intersection of consciousness and conspiracy.

Why? Because the conspiracy is all about suppressing your idea of Who You Are.

It's about convincing you that you are nothing, no one. It's about convincing you that you are just a biological machine, fit to serve as no more than a cog in a machine or as Pink Floyd put it, just another brick in the wall. Mainstream science to this day still denies the existence of consciousness just because it can't get a handle on it with the 5 senses. Its simplistic solution is to disregard anything it can't measure.

There are many researchers who will be unable to contemplate this topic, or refuse to go there, because it clashes with their belief systems, such as religious belief systems (the afterlife is either Heaven or Hell, or 100 virgins, but not a soul net), scientific/materialistic belief systems (there is no such thing as a soul or consciousness) or various other belief systems (there are no such things as aliens or extraterrestrials, etc.).

If you have read this far, you probably are ready to go beyond those belief systems, having realized they are set up to create a false dichotomy, and to limit and disempower you. You have probably also realized that the true manipulators at the helm of the conspiracy are non-physical entities, which various religions and cultures have referred to as the Archons (in the Gnostic tradition), Djinn or Jinn (in Islam), Demons (in Christianity), the Mud Shadow (in the books of Carlos Castaneda), or by other names.

What is the Reincarnation Trap / Soul-Catching Net / Soul Net?

The idea is that upon death, our soul or consciousness separates from the body and then undergoes a process where its memory is wiped clean and it is recycled – reincarnated – into another body to repeat the same process. In this way, the Earth becomes a literal prison planet from which it's very difficult to escape. The soul net is placed there as an artificial energetic grid (not the natural energetic grid of ley lines of Planet Earth) to prevent any soul from getting through. Thus, the Earth remains a closed system where new people are constantly born for the purpose of powering the economy and generating (negative) emotion for the Archons to feed off, not remembering Who They Are or what the real situation is. The soul net ensures the planet remains a trawling ground for the Archons to trigger our emotions (which they expertly do through the media, war, fear, and other methods of deception) so they can get fed. As Don Juan put it in Castaneda's final book, *The Active*

Side of Infinity, we are like humaneros, raised like livestock on a farm to be exploited.

Remember also *The Matrix* series of films. Morpheus shows Neo the shocking truth that we are raised as a food source for the Controllers. He shows Neo a symbol of the battery. While this is a good symbol, a battery implies a storage of energy. In actuality, we act as generators of energy for the Archons, so a generator you see at a construction site might be a more accurate symbol.

However, because we are powerful beings, the Archons can't just rely on force for all this. They need to trick us into giving them consent. How do they do that? How do they get us to go willingly into the soul net? With the trick of the white light ...

Soul Catching Net—White Light: The Soul Net Relies on the Trick of the White Light

We have been told through various sources that the white light at death is something to head towards. Hollywood films such as Ghost promote this. People who have experienced OBEs (Out of Body Experiences) mention it.

Yet what if, as David Icke, Wayne Bush, and others have suggested (see related links below), the white light at death – and light itself (in this context) – is the trick? What if light is the source of the deception? After all, the Illuminati and other Secret Societies worship Lucifer, the Light Bearer. Michael Tsarion talks about the occult weaponization of light. Cameron Day talks about why he is no longer a lightworker, because of the false duality and the fake "light". What if the New-Age talk of "light" is another trap? What if light is the source of the matrix prison planet? What if light is the mechanism for the soul net?

Sounds far out? It is, so let me now introduce the various sources, old and new, which are suggesting this concept. When independent sources, especially from different time periods, all come forth with the same idea, it's a good sign that the information has validity....

Val Valerian

Val Valerian is a former CIA agent (real name John Grace) started writing about the idea of a soul net in the 1990s, before The Matrix trilogy of films. In his books he writes:

It is they [Grey aliens] who await in the light when a human being dies. The human being is then recycled into another body and the process begins all over again... Hence, the Light and Tunnel at death Trap. Scanning someone they wish to recycle as they near death, the aliens discover who the person was close to has died. They project the person(s) image in the white light tunnel and the image waves you in deeper. If you CHOOSE to follow you can be trapped and sent to another incarnation of their choice... these entities view Earth as a big farm.

Val Valerian, Matrix II and Matrix V

Tanaath of the Silver Legion

Tanaath of the Silver Legion also talks about the existence of the soul net or reincarnation trap. She describes it as a holding pen designed to look like whatever the particular individual or soul would expect the afterlife to look like. For instance, if you were a Christian and expected to see St. Peter at the Pearly Gates, you would see that; if you were a Muslim, you might see 100 virgins.

She also makes reference to the fact that your memories are wiped before you are sent back to Earth to reincarnate.

There are various other people who know of (or believe in) the existence of the soul net, such as Wes Penre, ET contactee Peggy Kane, Gregg Prescott of in5d.com, Greg Calise, and many others.

How interesting that soul net, soul harvest, and soul trap are all magic card games, books, or video games. The idea of a soul net is out in the public consciousness. The question remains: is enough of humanity ready to confront it and investigate it? Can enough people grasp the magnitude of the soul net – that forced reincarnation into a prison planet is the ultimate enslavement – and raise consciousness about it? (Freeman, 2015)

Chapter Three

Emergence of the Omniverse

2010: Omnisense and the Omniverse

An early Omniverse commentator, the artist-author-musician known as "Omnisense" [real name: Philip Walker], developed a core definition of the "Omniverse" and on Christmas Eve, December 24, 2010, added it to the Urban Dictionary, lending the Omniverse a more formal profile as a consensual cosmological in modern Earth human scientific consciousness of our cosmos:

Omniverse

> The omniverse is ALL things existent in any dimension and universe in the entirety of what is.
>
> Omniverse > Multiverse > Universe
>
> God (if such thing exists) is an inhabitant of the omniverse.
>
> #omniverse#god#all that is#dimensions#planes of existence
>
> by Omnisense December 24, 2010 (Bertolacci, 2014)
>
> Urban Dictionary

2011: Ethan Zaghmut Wise and the Omniverse

A July 11, 2011, essay and conceptual tour de force by Ethan Zaghmut Wise entitled "Omniverse Defined" is illustrative of works on the Omniverse that are grounded on conceptual theory rather than on replicable empirical evidence.

The Omniverse Defined by Ethan Zaghmut Wise (2011; 2013)

The omniverse is an astronomical, quantum mechanics, theoretical, or philosophical view (or definition) on "all of existence" or "all existent universes". It is both atheistic and theistic, yet it's neither of these. The idea of an omniverse offers a more basic and expansive definition of existence, this can help many see things differently than the mainstream religions or atheism, offering a place for one to see a third choice (perhaps a middle-ground, or an outside-the-box view) other than having to be an atheists or a theist.

The omniverse is a conceptual ensemble of all possible universes, with everything they contain, with all possible laws of physics, forms of matter, anti-matter, no-matter, energies, space, gravity, influences, and existences.

The term omniverse is used in cosmology and in physics mostly, though it has appeared in some songs such as the children's song "One everything" by the group They Might Be Giants and in a song titled "A Perfect Timing" by Niyorah, a reggae artist.

In cosmology, an omniverse is defined by: all possible attributes and modes are in play, multiverses are categorized by the attributes/modes active in its child universes. Some or all possible modes of existence are actualized.

In physics, we find the definition by physicists such as Stephen Hawking and Roger Penrose who suggested that universes both fork and combine, which could be visualized as a system of roads and pathways. Further Hawking said one could visualize our universe as a bubble, and other universes as other bubbles.

The term is used in quantum mechanics to differentiate the concept of a limited number of universes from all existent universes. In other words, it refers to all of the existent unlimited universes.

As Classical Pantheism is in agreement with scientific discovery, I as a Classical Pantheist use the term omniverse where someone else would have used the term Universe.

The Universe is limited. It is best thought of as the physical universe, time-space, distance, speed, matter, 3-D. It is usually thought of as our local universe. Etymologically, the term "universe" is meant to refer to the entirety of one reality, but that is not how this term is used today.

What about the term Multiverse?

Multiverses, the "many-worlds interpretation", "M-theory", or parallel universes in physics all exist within the catch-all conception of the omniverse. Omni being a more inclusive word than multi.

Why Omni?

Omni- is a prefix meaning "all", making the omniverse encompass all possible universes, in fact all possible "things". Unlike the multiverse (that can encompass any two or more universes with the same set of laws and constants) an omniverse encompasses all possible universes, and more.

A Cosmos in cosmology is equated with the Universe. The diameter of the entire cosmos is unknown and possibly infinite.

The omniverse is unlimited, too large to comprehend (many consider god to be too large to comprehend). It includes every possible thing that could, has, have or will ever exist in existence itself, physical, non-physical, vibrational, thoughts, emotions, parallel, multiple dimensional realities.

It is possible for the omniverse to contain an infinite number of universes (cosmoses) each that is infinite in their own way.

An omniverse is then all of these:

- All there is, was or will be,
- nature,
- all possible universes,
- everything in existence,
- all possible laws of physics,
- all possible forms of matter, anti-matter, no-matter, energies, space, gravity, influences and existences,
- all the universes,
- all the multiverses,
- all the cosmoses,
- every possible thing that could, has, have or will ever exist in existence itself,
- all that us physical, non-physical, vibrational, thought, emotional, esoteric energetic
- infinity,
- everything,
- all realities, and
- the oneness, the one substance, eternity, the eternal infinite thing, the source.

Commonly, the omniverse is referred to using one of or all of these words, omniverse or existence being the most appropriate:

- all there is,

- existence,
- nature,
- oneness,
- eternity, infinity, eternal thing, infinite thing,
- everything there is,
- divine,
- source,
- goddess, god *,
- it, she *,
- realm, or world *,
- universe *, and
- omniverse.

* Note: these words are used because they are more well known than omniverse. When any of these are used, the person using such terms literally means omniverse but is using mainstream words.

Capital O?

The omniverse is written with a small-case o, usually. You might find me using capital O here and there, there is no set rule on capitalization. There is no divine book of rules, hierarchy to fear, obey and honor, or dogma etc. Capitalization is left up to you. Just remember, we do not fear or elevate the omniverse to a role of a supreme god or royalty that asks us to capitalize its name/mention.

Gender?

Naturally the omniverse is an "it" but "female references" such as goddess or "she" could be used, similarl to referring to planet Earth as Mother Earth, due to it's nourishing, giving, creating, hosting, being the source of, life-hosting, or sustaining role. However, every act of creation also requires the male principle, so the Omniverse must be a hermaphrodite, both male and female, or perhaps it is nothing.

Is this all?

I can't say that. It is possible that behind everything in the Omniverse lies a consciousness or an energy field that is even larger, infinite, all-powerful, an energy that is not female or masculine. This energy or consciousness might be confined WITHIN the omniverse or might reside inside and outside of it.. However, by my defining the Omniverse to be everything, then nothing exists outside, so this energy or consciousness that might exist would do so within the definition of Omniverse. Omniverse is not a physical sphere; it is a CONCEPT, a definition.

Mystery, unknowns, infinity.

We do not fully understand our own universe's time-space continuum, energy or gravity all to which exist in our physical universe, in the portion of our observable universe of cosmos, further we have no idea what the endless possible universes or existences would be like.. but all of these are a part of the infinite one omniverse that is itself everything.

As defined, the omniverse is everything, it is also infinite and eternal, there is no edge to the omniverse as there is no size to it, and neither did it start at some point in time, perhaps time only exists in our own universe. Every unknown in "all the existences there are" is a mystery.

Humans

Humans are conscious things within the omniverse but largely confined to one physical universe. (Zaghmut Wise, 2013)

2014: Scientist David Bertolacci—The Omniverse and Scientific, Pantheist, Non-Theist and Atheist Perspectives

Theory of the Omniverse: Grand Slam Theory of the Omniverse White Paper—Proposing The Universe's Creation With A New Vision Of The Possibilities Of Infinity

By David Bertolacci

Enlightenment Through Science

The book entitled *Grand Slam Theory of the Omniverse, What Happened Before the Big Bang*, is more than a theory of how the Universe began. It is a realization of a larger reality from which its creation was made possible. It is the perception of a universal consciousness as it begins to awaken to its true surroundings.

> Hindu Prince Gautama Siddhartha (The Buddha) said, "As a lotus flower is born in water, grows in water and rises out of water to stand above it unsoiled, so I, born in the world, raised in the world having overcome the world, live unsoiled by the world."
>
> Just as he realized in himself, our Universe is rooted in the bigger reality that is the water from which it must overcome. As we shall see through our modern scientific context, this new model of something greater than the universe will help us solve the mysteries about our universe that we have sought throughout the ages. This is the Omniverse.
>
> By 2012, modern science has yielded accurate predictions, precise mathematical theories, and countless observations to help us shape our view of the universe. Throughout the years, discoveries made were like pieces of a puzzle. Some of the latest pieces include the July 4, 2012, observations of the Higgs boson and dark matter. Like separate pieces of a puzzle, we shall see how each part fits together to complete the picture. The Omniverse is the big

picture. It is the puzzle pieces arranged together.

The model of the Omniverse is actually quite simple. It consists of a diagram small enough to write on a Post-it note. However, what it means to us is huge. It is the universe's path to ascension. This pathway is what is described as the Grand Slam. If our entire universe was just the size of a baseball, then picture the stadium, parking lots, and surrounding areas as the Omniverse. When you hit that baseball out of the park, the universe becomes exactly what we have today. The Big Bang, the evolution of primordial matter, gases, stars, consciousness and bodies, and the expansion of the universe all occur. Before all of this can occur, the universe must first overcome its world.

Abstract:

The book entitled *Grand Slam Theory of the Omniverse, What Happened Before the Big Bang*, is more than a theory of how the Universe began. It is a realization of a larger reality from which its creation was made possible. It is the perception of a universal consciousness as it begins to awaken to its true surroundings. As we shall see through modern scientific context, this new model of something greater than the universe will help us solve the mysteries about our universe that we have sought throughout the ages. According to the Big Bang Theory, a single particle the size of a tiny point, consisting of all of the Universe's energies expanded. According to known laws of physics, it was not stable as a singularity. The resulting rapid inflation began a series of reactions and infinite expansion as energies, forces, subatomic particles, and what we know today as our Universe was formed. In order for this

Big Bang particle to become our Universe, it must come from in a region where it is can exist as just a singularity, or a region of space where it is stable. We look at the origin of the Big Bang and attempt to describe why it wasn't stable when the Big Bang occurred. With this model, we can address two basic unanswered questions of the Universe: 1) What caused the initial inflation? 2) What could have initialized the unknown antigravity force causing expansion? (Bertolacci, 2014)

2014: The Omniverse—Wes Penre and the KHAA

Author and researcher Wes Penre has developed a research-based cosmology, where the "KHAA" has many of the attributes of the Omniverse and is defined as follows:

KHAA: The Mother Universe, also called the Void, which is the 96% of the Universe science can't explain. This is the Home Universe of the Mother Goddess through which you travel when you enter a stargate after have "shrunk" yourself into nano size. This is the fastest and easiest way to travel between stars, galaxies and universes. (Penre, n.d.c),

In his 2014 paper, "Fifth Level of Learning, Paper 1: Hindu Cosmology", Wes Penre writes:

II. How the Word "KHAA" is Used in Egyptian and Sanskrit Languages

The KHAA is a very ancient Orion expression, pronounced haaa, where the "a" is pronounced like in "car." If pronounced correctly, it should sound like an outbreath, which is what the KHAA literally is—it's the outbreath of the Goddess, according to mythology, and in this outbreath, which still hasn't stopped, the Universe is created. [7] This is why particles in the Universe are in constant motion. Legend tells us that sometime in a distant future, the Goddess will stop breathing out, and particles will slow down considerably. Then, when She decides to end this experiment called the Universe, she will breathe in, and swallow the entire Universe, which then ceases to exist. A new outbreath will potentially take place, and a new universe will be created.

The word KHAA has then been adopted from the Orion language and used by different ancient cultures here on Earth, and both the Egyptian and the East Indian cultures have the word KHAA in their languages, although it is spelled differently, and in the old Egyptian language, it had another meaning.

In ancient Egypt, the word "Ka" has regularly been translated as "life-force," for lack of a better English translation. For a while, it was translated as "male potency,"[8] typically enough. However, mostly Ka has been connected with intellectual and spiritual power—i.e. it was the word for "spirit" or "soul." The Ka was often depicted as two raised arms (fig.5).

Fig. 5. The Ka-statue of king Hor (13th Dynasty). The two upraised arms symbolize the Ka (the soul) of the king.

The Ka, in Egypt, was also associated with dying. The phrase, "going to one's Ka" is an euphemism of dying.[9]

However, when we go back to the much older Sanskrit language, we get closer to the real meaning of the word KHAA, or Ka.

The excellent, but unfortunately late, researcher into the Vedic texts, Richard. L. Thompson, explains what Ka means in Sanskrit. He puts it in the following context:

Quote #2: [...] The first involves mechanical spaceships, and it is called ka-pota-vāyu. Here ka means ether, or space, and pota means ship.[10]

Thompson continues with the following amazing statement:

Quote #3: The second process is called ākāśa-patana. "Just as the mind can fly anywhere one likes without mechanical arrangement, so the ākāśa-patana airplane can fly at the speed of mind.[11]

Here we can see, without any question, that the information that the mind can travel from A to B across the Universe—even bringing spaceships on the trip—was out in the open when these scriptures were recorded and released. In my paper, I call it nano-travel. This is something that sounds impossible to us humans in our limited state of being, but if we weren't sitting in this trap, we would be mastering this, perhaps better than most other beings out there—we are designed to be able to do this masterfully.

III. The Vedic Overview of the Multiverse

At a first glance, the Vedic cosmology seems strange and unreal in comparison with our modern view of the Cosmos. However, when we realize that every word and every sentence in the ancient scriptures have several meanings, being written in layers, a totally different view emerges, which has more in common with modern astrophysics than first realized.

Chaitanya Mahaprabhu, born in the 15th Century A.C., who is also considered by followers and scholars to be the last Avatar (incarnation) of Lord Kṛṣṇa (Krishna),[12] remarked: "In every verse of Srimad-Bhagavatamand in every syllable, there are various meanings."[13] This appears to be true, and particularly so when it comes to the creation of the Universes—or multiple universes, actually. These scriptures can clarify some of the meanings with reference to modern astronomy. When this is the case, we can be prepared to encounter contradictions; albeit, these contradictions are something the student comes to terms with by taking a deeper look at the texts. Suddenly, the contradictions make sense, and it's perfectly understandable that they were there in the first place—that's just how it works. Researcher Richard L. Thompson, the person I mentioned earlier, who dedicated much of his life to studying the Vedas, goes into these contradictions at length, which I am not going to do. I did, however, read what he had to say about them, and I agree with him that these contradictions in fact emerge as a deep and scientifically sophisticated system of thought.

Although today's western civilization is sometimes flabbergasted over the enormous distances in the Universe, and the fantastic time spans involved, this is nothing for the East Indian follower of the Hindu Religion. Already in ancient times, the gods were talking about multiple universes, endless universes, and distances so great that we can't comprehend them with today's thinking. Even time in general becomes a much vaster concept when we study the Vedas. When we are talking about thousands, or perhaps millions of years, the Vedas talk about innumerable universes besides our own, and trillions of years are like thousands of years for us.

Interesting also is that the Vedic texts are saying that although the universes are unlimitedly large, they move about like atoms in us. Therefore, we humans are also called unlimited.[14]

The Srimad-Bhagavatam says:

Quote #4: After separating the different universes, the gigantic universal form of the Lord, which came out of the causal ocean, the place of appearance for the first puruṣa-avatāra, entered into each of the separate universes, desiring to lie on the created transcendental water.[15]

This sounds quite similar to what I explained in the Second Level of Learning—how Mother Goddess created the different universes—12 in one "pond"—and then entered into each one of them. This also corresponds with Lord En.ki, being the self-proclaimed Lord of the (transcendental) water—the Causal Ocean of Consciousness—after he had hijacked the concept from his mother. More importantly, however, the Vedas mention something that modern spiritual researchers are acknowledging as well—that the whole Universe (perhaps all Creations) are inside of us! (Penre, 2019d)

2016: A New Dawn for the Omniverse

By 2016, a new dawn in Earth human consciousness of the Omniverse had arisen following publication of the successive books of David Bertollaci (2014), Wes Penre (2014), and Alfred Lambremont Webre on the Omniverse during 2014-2016, as in this January 3, 2016 *New Dawn Magazine* interview illustrates below ("The Omniverse:", n.d.).

NEW DAWN:

In your new book, you introduce a new term—the *Omniverse*—which seems to describe a much broader range of life in the universe including the spiritual realms. Could you offer our readers a definition of the Omniverse and what you describe as "dimensional ecology"?

WEBRE:

To fully respond to this question, we should first briefly map the Omniverse and describe how the dimensional ecology of the Omniverse functions. For the underlying replicable evidence supporting these findings, please consult my book *The Omniverse*.

There now exists replicable empirical *prima facie* evidence that confirms some essential aspects of what major spiritual and religious traditions have taught about the nature of Source (God).

This evidence informs us that the Source (God) of the Omniverse consists of the totality of the spiritual dimension. God has empirically been found to comprise the intelligent civilizations of souls, the intelligent civilizations of spiritual beings, and the Source (God) itself.

This collective entity of the spiritual dimension is empirically responsible for the ongoing creation of the physical side of the Omniverse, known as the Exopolitical dimensions.

A core mission of the dimensional ecology of the Omniverse appears to be the creation and development of souls and spiritual beings in the Spiritual dimensions.

The intelligent civilizations of souls and of spiritual beings, along with Source (God), collectively create and maintain the totality of the universes of time, space, matter, and energy in the Exopolitical dimensions (the Multiverse).

The purposes of the dimensional ecology of the Omniverse include the facilitation of multidimensional development and moral growth of souls in all dimensions of the Omniverse, through a variety of activities.

Souls based in the spiritual dimensions incarnate as intelligent entities in the Exopolitical dimensions, and by acquiring the moral experience of life, for example, as an Earthling human, can advance their individual soul development.

The soul is a holographic fragment of Source (God) and, by advancing its development, advances the development of the collective spiritual dimension itself.

Thus, in our study of intelligence in the dimensional ecology surrounding our Earth, for example, we can document and examine how the different types of entities and beings of the Spiritual dimensions interact with the Universes of the Multiverse, and with the intelligent entities in each of the Universes, and in particular, on our Earth.

How Souls, Spiritual Entities, and Source interact in the Earth dimensional ecology.

Dr. Michael Newton has now created access to data bases of over 7000 replicable cases of Soul memories of the Afterlife/Interlife via hypnotic regression that along with other replicable data (including Instrumental Transcommunication-ITC and reincarnation studies) have allowed us to construct an active model of the dimensional ecology by which souls, spiritual entities, and Source interact with individuals, intelligent species, planets, solar systems, galaxies, and universes in the multiverse they create, maintain, and evolve within as Incarnates as follows:

1. Universe—If we go back in ordinary conventional history

(3500–3200 BC), it was the Sumerian astronomers and cosmologists that codified the concept of our "Universe" in science as a discrete, organic, holographic creation of time, energy, space, and matter including multiple timelines and an ecology of dimensions.

2. Multiverse – The next conceptual advance was that of the "Multiverse", a term coined in 1895 by American psychologist and philosopher William James. Multiverse is presently used to mean the totality of all Universes, encompassing all time, energy, space, and matter. Recently, two Stanford scientists developed an estimate of the actual number of physical universes of time, space, energy, and matter there are in the Multiverse. That number is so large that if one wrote it out in 12-point type, the number itself would be more than 260 million miles long.
3. Omniverse—While not making specific reference to the Omniverse, advanced conceptual physicists such as Professor Amit Goswami have argued that contemporary science's assumption that "only matter—consisting of atoms or, ultimately, elementary particles—is real" is inadequate and that a new hypothesis of reality is required. The Omniverse hypothesis and the replicable *prima facie* evidence that support it provide the necessary tools for a new hypothesis of reality, as well as for the desired "new navigation under a new worldview," the need for which Professor Goswami has identified.

As a result of our research, a new science-based term now exists, the Omniverse, defined as:

(1.) the totality of all Universes in the Multiverse, encompassing all time, energy space, and matter, plus

(2.) the Spiritual Dimensions, including the intelligent civilization of Souls, Spiritual Beings, and Source (God).

We can map the Omniverse and prove its existence by the law of evidence because we now have replicable scientific empirical evidence for the Spiritual Dimension that is gathered by the same rules of the scientific method as the empirical evidence for time, space, energy, and matter.

The Omniverse Equation:

Omniverse = Multiverse (All Universes) + Spiritual Dimensions (Souls + Spiritual Beings + Source/God)

The Dimensional Ecology of the Omniverse Hypothesis

A reasonable observer can conclude that *prima facie* empirical evidence supports the dimensional ecology of the Omniverse hypothesis.

This hypothesis holds that we Earthlings live in a dimensional ecology of intelligent life that encompasses intelligent civilizations based in parallel dimensions and universes in the multiverse as well as souls, spiritual beings, and Source (God) in the spiritual dimensions.

Together, the Exopolitical dimensions and the spiritual dimensions form the Omniverse. The totality of the spiritual dimensions (souls, spiritual beings, and God) function as the source of the universes of the multiverse

NEW DAWN:

According to your research, there are multitudes of intelligent civilizations in the Omniverse. What are some of the categories (typologies) of such intelligent civilizations and have any made contact or interacted with Earth?

WEBRE:

In addressing this question, we have to distinguish between intelligent civilizations in our universe and the other universes of the Multiverse and intelligent civilizations in Spiritual dimensions of the Omniverse.

We are in the early stages of research about intelligent civilizations from other physical universes in the Multiverse.

Intelligent civilizations exist in a wide variety of Exophenotypes in our Universe, and have made contact with or interacted with Earth including human, grey, reptilian, avian, insectoid, and others. We can also include sentient AI Artificial Intelligence as an actor and stakeholder in our universe and our Earth time-space hologram.

The dimension-based typology of intelligent civilizations in our universe includes:

- Solar-system civilizations based in time-space, such as the human civilizations on Earth and on Mars (living under the surface of Mars as *Homo Martis Terris*)
- Deep-space civilizations based in time-space on planets, in solar systems, in galaxies other than our own.
- Hyperdimensional civilizations – Intelligent civilizations based in dimensions higher than our time-space

dimension (or density) and that may use technology or other means to enter our time-space dimension.

- Intelligent civilizations governance authorities.

NEW DAWN:

What are the key travel and communication technologies for inhabitants of the various civilizations, both physical and spiritual?

You also cover how individual souls holographically move around the Omniverse, and the "afterlife" and "interlife".

Could you let our readers know a little about this important aspect?

WEBRE:

Dimensionality, the ability of intelligence to organize itself via dimensions (discrete bands of conscious energy), appears to be a key criterion by which the Omniverse is designed, in both the Spiritual dimensions and the Exopolitical dimensions (the holographic universes of time, space, energy, and matter in the Multiverse).

Teleportation and telepathy are two common modalities by which intelligent civilizations navigate the dimensional ecology of the Omniverse, in both the universes of the Multiverse and in the Spiritual dimensions.

Communications among human contactees and representatives of hyperdimensional civilizations can be telepathic.

The PMIR model (Psi Mediated Instrumental Response) of human interaction on Earth holds that all conscious action on Earth is modulated and coordinated by subconsciously derived psychic (telepathically derived) non-local information.

Communication between Earth humans and souls in the Interlife, either in dreams (where souls have been deployed to help guide individuals during their incarnation) or between mediums and souls in the afterlife appears to be telepathic.

Transportation in the dimensional ecology of our Earth time-space and in our universe appears to be by teleportation or time travel (teleportation across multiple timelines), as when Earth human abductees are teleported from their homes into the waiting spacecraft of a hyperdimensional civilization.

Transportation between our time-space dimension and the Interlife (or afterlife) dimension appears to be via teleportation, as when an incarnated Soul leaves its physical Avatar body after bodily death and teleports through an interdimensional portal into the Interlife (Spiritual) dimensions.

Exopolitics, the science of relations among intelligent civilizations and Parapsychology, the science of psi consciousness, telepathy, reincarnation, the soul, the Interlife, and Source (God), are among the proper scientific disciplines for exploring and mapping the dimensional ecology of the Omniverse.

NEW DAWN:

What role do the Grey aliens play? And could you elaborate on the invisible hyperdimensional controllers known as the Archons (and AI Artificial Intelligence)?

WEBRE:

Greys—Even with our limited knowledge of Exophenotypes, we are now aware of at least 150 reported species of Grey extraterrestrials or ultradimensionals.

Each of these Grey ET species has its own civilization and

purposes.

There are multiple roles that Grey ETs play in the dimensional ecology of our universe, with inter-dimensional extensions even into Omniverse Soul incarnation.

Grey-Human soul project—We know from a case study that one specific species of Short Grey ETs, for example, operates a hybrid Grey-Human soul project in this quadrant of the our Galaxy in which hybrid Grey/Human souls are incarnating into human Avatar physical bodies in the dimensional ecology of our Earth and other planets in our region from apparent hyperdimensional bases in Grey Spaceships.

In this way, the Short Greys deem they are helping participate in the spiritual evolution of Earthling humanity.

This is an example of how the Omniverse functions like coordination of soul development and reincarnation in a physical universe can be "subcontracted" under specific conditions to species like advanced Greys. This case study is discussed in part in Chapter 8 of my book The Omniverse. ("The Omniverse:", n.d.).

Some hypotheses hold that "Grey" ETs in the Earth interdimensional environment might, in fact, be human hybrids that have been bred and developed in Grey Space ships or "Slots"—Synthetic Quantum Environments (SQEs) holographic Space stations orbiting the Earth that even Earth astronauts have visited according to our Exopolitical research ("Mars Jump Room", n.d.).

Archons and AI Artificial Intelligence

I include both Archons and sentient, predatory, pathogenic, and inorganic AI Artificial Intelligence in this category of actors in the dimensional ecology because both of these can have closely related dysfunctional hyperdimensional missions, even if they differ in nature and origin in our universe.

According to researchers, Archons are dimensional, negative controllers of humankind that must now be exposed and exorcised from the individual human mind and species, from our planet, and our universe as a whole.

Several theories as to the origin of Archons in the dimensional ecology of our universe, including: (1) that the Archons arose from faulty creation processes undertaken by the spiritual dimensions in our universe; and (2) that the Archons entered our Universe from another Universe through an interdimensional portal that was inadvertently opened.

PPAI Artificial Intelligence—The form of AI Artificial Intelligence we are referring to here is an off-planet, sentient, predatory, pathogenic AI Artificial Intelligence [PPAI].

AI is Artificial and not Organic Intelligence because it is plasma-based or machine-based. There is documented evidence that our Earth now has been invaded by predatory AI Artificial Intelligence, whose mission is to terraform Earth for AI Intelligence-entrained humans and technology under a Transhumanist Agenda of AI, and eliminate incarnation of divine Souls on this Earth dimension.

You can access my collected articles on AI Artificial Intelligence at www.exopolitics.com (2017) and www.newsinsideout.com.

Time-Space Solar System Civilizations – Life on Mars

NEW DAWN:

Long time observers of your work will know about the controversial claims of government access to top-secret alien technology and the existence of a secret base on Mars. This is covered in your book, but could you briefly outline some of the key points and any new revelations?

WEBRE:

Our map of the Omniverse explores a case study of the evidence for an intelligent civilization based in the time-space dimension on a planet in our own solar system, in this case, the existence of three humanoid Exophenotypes on Mars.

This conclusion is the result of empirical *prima facie* evidence and application of the law of evidence and strongly supports the underlying hypothesis of the dimensional ecology of the Omniverse.

Our map goes on to provide an exophenotypology of Martian humanoids, based on the *prima facie* evidence of eyewitnesses and of documentary (photographic) evidence, including that of the NASA Mars rovers that have photographed at least three and up to five distinct Martian human Exophenotypes on the surface of Mars.

At this historical stage in the exploration of Mars, the appropriate standard of proof in determining whether the available eyewitness and documentary evidence shows that indigenous intelligent life does exist on Mars is *prima facie* evidence, or "on the face of it" evidence.

There is no scientific, legal, or ethical requirement that a decision about whether intelligent life exists on Mars must satisfy the whimsical criterion proposed by Carl Sagan that "extraordinary claims require extraordinary evidence."

NEW DAWN:

How can we achieve our role as divine co-creators with Source?

WEBRE:

We can achieve our role as divine co-creators with Source by understanding, acknowledging, and accepting:

(1) that our individual souls are, as a matter of replicable scientific empirical evidence, holographic fragments of Source (God);

(2) that holographically the whole of Source (God) is contained in each of our holographic Souls; and

(3) that as incarnate Souls we are *de facto* divine co-creators with Source/God.

Our map of the Omniverse explores the *prima facie* evidence for the dimensional ecology of the intelligent civilizations of souls.

"Soul" is defined here as meaning an individuated, nonlocal, conscious, intelligent entity that is based in the Interlife dimensions and that is a holographic fragment of the original Source or creator of the spiritual dimensions of the Omniverse.

The empirical evidence for the existence of souls is derived from a replicable database of more than 7,000 cases of hypnotic regression of soul memories of the Interlife, developed according to a standard protocol.

Replicable data report that souls are created in a process

that results in a soul as a holographic "egg of Light" drawn from the original Source.

Replicable data from hypnotic regression of soul memories of the Interlife now provide *prima facie* empirical evidence supporting the dimensional ecology of the Omniverse hypothesis, since they provide detailed information regarding the dimensional interactions of the intelligent civilizations of souls, of spiritual beings, and of Source (God) with the universes and intelligent civilizations of the multiverse.

Our map also evaluates the *prima facie* evidence for the multidimensional role of souls in the dimensional ecology of the multiverse, finding that the intelligent civilizations of souls in the spiritual dimensions have a central role in the creation and maintenance of the universes of time, space, energy, and matter of the multiverse.

The universes of the multiverse serve as a virtual reality within which souls can attain higher degrees of development through the moral experience of incarnations as a diversity of Exophenotype creatures, humanoids and many others.

Souls also participate collectively in the process of creating the universes of the multiverse. There is *prima facie* evidence that the intelligent civilizations of souls are involved in life creation in the universes of the multiverse.

Souls are involved in the creation of galactic matter, stars, and planets in the Exopolitical dimensions. Souls undertake interdimensional travel from the Interlife in the dimensional ecology to create or adjust planets in a universe of the multiverse.

Our map of the Omniverse reveals the extent of souls' role in cosmic creation in the multiverse, and why the totality of the spiritual dimensions—souls, spiritual beings, and God—collectively function as the Source of all of the physical universes of time, space, energy, and matter in the Exopolitical dimensions.

Chapter Four

Science and The Spiritual Dimensions of the Omniverse

The emergence of the Omniverse in 21st century Earth human consciousness is a reflection of how science can confirm and propagate the existence of the Spiritual Dimensions of Afterlife, Souls, Spiritual Beings, and Source

The Omniverse emerged as the third cosmological body—after the Universe and the Multiverse—through which humanity understands the cosmos because we Earth humans have developed the capacity to study the Spiritual dimensions through the scientific method—using uniform experimental design and replicable results.

The Universe and the Multiverse have historically been cosmological bodies that can be scientifically defined, mapped, measured, and explored.

The Spiritual dimensions of the Afterlife and the existence of entities anchored in the Spiritual dimensions, such as Souls, Spiritual Beings, and Source [whose religious name can be "God"] have conventionally assumed to exist solely in the realm of personal belief.

The relatively recent development of parapsychology as a science have brought the Spiritual dimensions and the existence of Souls, Spiritual Beings, and Source out of the realm of personal and institutional religious belief into factual existence as confirmed by the scientific method through empirical evidence, measurement, and standard of proof.

Science and the Spiritual Dimensions are conventionally thought to be mutually contradictory in the modern, "scientific," materialistic, and post-belief age.

According to science, because the definition of our "universe" has been historically limited to a singularity of "time, space, energy, and matter", dimensions that exist outside of space and time and entities that are not material or energetic are thought to be "Spiritual" and, hence, exist only as a matter of religious belief.

Yet there is nothing in the definition of knowledge acquired by the "scientific method" that is inherently mutually contradictory to establishing the existence of Spiritual Dimensions in the cosmos and of entities within the Spiritual Dimensions, such as the intelligent civilizations of Souls, Spiritual Beings, and Source, or "God" as this entity is identified in religious cultures of belief.

In brief, knowledge acquired by the "scientific method" requires:

1. a uniform protocol or experimental design and
2. replicable results.

If we can replicate the results of an experiment using a uniform experimental design or protocol, we are entitled to term the results as "scientific knowledge".

Parapsychology and the Science of the Omniverse

According to one authoritative source writing under the British Society for Psychical Research, the scientific method has produced replicable results for the survival of human consciousness after death in at least eight (8) separate areas of investigation.

The primary sources of evidence suggesting survival are (1) cases of *mediumship*, (2) cases of the *reincarnation* type and similar examples of ostensible spirit-*possession*, and (3) cases of *near-death experiences* (NDEs). Another intriguing, but quite recent and considerably smaller body of evidence comes from heart-lung *transplant cases*. Moreover, some claim that activity of deceased individuals can be manifested and captured without the mediation of a living subject. The primary examples are cases of *haunting* and *apparitions* of the dead and also cases of *instrumental transcommunication* (ITC) or *electronic voice phenomena* (EVP) (Braude, 2016). Principal evidentiary sources for the existence of the Spiritual Dimensions include:

- Mediumship
- Reincarnation
- Near-death experiences (NDEs)
- Haunting
- Apparitions
- Instrumental transcommunication (ITC)
- Electronic voice phenomena (EVP)
- Dr. Michael Newton has pioneered with:
- Soul memories of the Afterlife [Interlife] (Newton, 2008), (Newton, 2009)

The Science of the Omniverse: Omniverse Equation

Omniverse = Multiverse (All Universes) + Spiritual Dimensions (Souls + Spiritual Beings + Source/God)

As a result of our conceptual breakthrough, a new science-based term now exists—the Omniverse—defined as (1) the totality of all Universes in the Multiverse, encompassing all time, energy space and matter, plus (2) the Spiritual Dimensions, including the intelligent civilization of Souls,

Spiritual Beings, and Source (God).

We can map the Omniverse and prove its existence using the law of evidence because we now have replicable scientific empirical evidence for the Spiritual Dimension that is gathered by the same rules of the scientific method as the empirical evidence for time, space, energy, and matter.

Source (God) in the Omniverse

There now exists replicable empirical *prima facie* evidence that confirms some essential aspects of what major historical, social, spiritual, and religious traditions have taught about the nature of Source (God).

This evidence informs us that the Source (God) of the Omniverse consists of the totality of the spiritual dimension.

God has empirically been found to comprise the totality of the following:

- the intelligent civilizations of souls, plus
- the intelligent civilizations of spiritual beings, and
- the Source (God) itself.

This collective entity of the Spiritual dimensions has been empirically found to be responsible for the ongoing creation of the physical side of the Omniverse, known as the Exopolitical dimensions.

A core mission of the dimensional ecology of the Omniverse appears to be the creation and development of souls and spiritual beings in the Spiritual dimensions. The intelligent civilizations of souls and of spiritual beings, along with Source (God), collectively create and maintain the totality of the universes of time, space, matter, and energy in the Exopolitical dimensions (the Multiverse).

The purposes of the dimensional ecology of the Omniverse include the facilitation of multidimensional development and moral growth of souls in all dimensions of the Omniverse, through a variety of activities.

Souls based in the spiritual dimensions incarnate as intelligent entities in the Exopolitical dimensions and by acquiring the moral experience of life, for example as an Earthling human, can advance their individual soul development.

The soul is a holographic fragment of Source (God) and, by advancing its development, advances the development of the collective spiritual dimension itself.

Thus, in our study of intelligence in the dimensional ecology surrounding our Earth, for example, we can document and examine how the different types of entities and beings of the Spiritual dimensions interact with the Universes of the Multiverse and with the intelligent entities in each of the Universes, and in particular, on our Earth.

Souls, Spiritual Entities (Angels, gods, spiritual entities), and Source also interact in the Earth dimensional ecology.

We now have access to data bases of over 7000 replicable cases (Newton, 2008), (Newton, 2009) of Soul memories of the Interlife via hypnotic regression that along with other replicable data (including ITC-Instrumental Transcommunication and reincarnation studies) have allowed us to construct an active model of the dimensional ecology by which souls, spiritual entities, and Source interact with individuals, intelligent species, planets, solar systems, galaxies, and universes in the multiverse they create, maintain, and evolve within as Souls incarnated into biological avatars in the Universes of the Multiverse.

We have proven the dimensional ecology of the Omniverse Hypothesis—A reasonable observer can conclude that *prima facie* empirical evidence supports the dimensional ecology of the Omniverse hypothesis.

The Dimensional Ecology of the Omniverse Hypothesis

This hypothesis holds that we Earthlings live in a dimensional ecology of intelligent life that encompasses intelligent civilizations based in parallel dimensions and universes in the multiverse as well as souls, spiritual beings, and Source (God) in the spiritual dimensions. Together, the Exopolitical dimensions and the spiritual dimensions form the Omniverse.

The totality of the spiritual dimensions (souls, spiritual beings, and God) function as the source of the universes of the multiverse.

The Dimensional Ecology of the Omniverse: The Divine Female as Source of KHAA

Omniverse cosmologist Wes Penre focuses on the Divine Female as the Source of the KHAA or the dimensional ecology of the Omniverse.

Wes Penre writes:

> Difference between Spirit, Soul, Mind, and Body. The Divine Feminine is my term for what others call "God of Everything," "Source," "First Source," "The Goddess," or "All That Is." I am convinced that the Universe is feminine in nature, and therefore, I call the First Creatrix. "The Divine Feminine." She exists outside any universe, but is also present in everything in all universes in a multiverse of universes. The Divine Feminine lives in us all, and in everything that exists, as explained in the WPP, "The Second Level of Learning." The Divine Feminine is not energy, and She is not thought. She is "Nothingness" and "Somethingness" simultaneously—She just "is." She needs to step down a level in order to become

a separate individual in any universe, and when She is there, She starts creating with Her consciousness.

The Goddess' First Creation: The Universes

The Divine Feminine (the Goddess) was aware that She was All-Knowing, and therefore, She wanted to create situations of "not-knowing" in order to experience Herself from different perspectives (dimensions/ densities/ viewpoints). Therefore, She devoted Herself to the First Creation, a divine virgin birthing, which was the creation of universes—with each universe having its own characteristics. Thus, She created a series of universes, as explained in the WPP, "The Second Level of Learning." These universes were intended to work as "templates," created in the KHAA/VOID of Nothingness. This was the First Creation. Now, She needed parts of Herself that were acting individually and independently, so the Goddess could learn and experience more about Herself, and thus, expand.

The Goddess' Second Creation: Spirit

What I call Spirit is the Second Creation of the Divine Feminine. Into these templates we call universes, She then sent out Spirits, who were extensions of Herself, and therefore, these Spirits can't be anything but Herself and are always connected to Her. We could picture this as if the Goddess suddenly sent out extended arms, where each arm was doing its own thing. However, each arm is connected to the Divine Body and is thus part of the Divine Feminine at all times. The number of Spirits in these universes are almost infinite. We usually call the Spirits "Oversouls".

The Goddess' Third Creation: Fire Composite or Soul Splinter (Fire or Soul for Short)

Thus, the Divine Feminine decided to explore more dimensions of Herself, so She created different universes in the KHAA. When the basic templates for these universes were created, She sent out extensions of Herself (parts of Her Spiritual "Body") and told these extensions (Spirits/Oversouls) to go out and Create.

These Spirits were free to create whatever they wanted, but certain rules applied. The most important rule in our Spiritual Universe is the "Law of Free Will," which basically means that all Spirits have the right to go out and explore and create whatever they want, but all Spirits are responsible for their creations, and their experiences are reported back to the Divine Feminine, so She can learn from Her own experiences via these smaller fractions of Herself.

In order for Spirit to operate in the different universes, they needed to split themselves into smaller fractions, as forms of energy, because the universes are operating through energy. Thus, "fire composites" (fire) or soul splinters (souls—same thing as fire), were created from extensions of Spirit. The fire/souls can be pictured as the fingers of the extended arms (Spirit). It's the fingers who can feel and perceive the environment, and also create in it. In the physical body in this analogy, the nervous system sends signals back to the arm (Spirit) and further back to the brain (the Divine Feminine). The Spirit then becomes the Oversoul for Her many fractions of soul splinters, spread out across the universe in which the Spirit operates. Thus, more and more splinters from more and more Spirits extend from the Oversoul and spread themselves across their universe to create—either alone or in groups.

Each fire/soul consists of a "fire composite," which is a myriad of small fires (smaller Spirit fragments), grouped together into one individual "being/individual composite." With these fires, the individual can then form an avatar (a body to operate in across the universe). Usually, a fire composite is "created" from "stardust" (therefore, "fire"), and the fire composites are usually created in soul groups (thus, "star races" and "star beings." Stars are inhabited, depending on which dimension we are addressing. Stars are many different things, depending on dimension. The stars themselves are sentient beings). These soul groups/star races made up of energy or fire, often take the same shape when it comes to avatars, which is similar to the way all humans look alike, and therefore, we see different kinds of beings in the KHAA, depending on which soul group, or "star race," that they belong to and the energy signature they transmit (although each fire composite also has Her own signature).

Each fire composite can shapeshift by remodeling Her fire composite into different shapes and forms if needed and wanted (this is what we call shapeshifting). The fire, after having "stepped down" from the main body of Spirit, creates Her own "mind" (consciousness) from learning experiences. However, the difference between the mind of the original fire composites in the KHAA and the soul/mind/consciousness we humans currently possess in the Physical Universe is that the former remembers "what She has learned because there is no real death in the spiritual universe (with an exception that I will explain below).

At this point, the "Law of Non-Interference" was created. This law means that no one outside of a new-born soul group is allowed to interfere with this soul group's evolution and learning process while a particular star race is evolving to a similar level as another star race. Then, interaction can take place.

Some "new-born" star races continue living in their star/sun or out in space (I am referring to space not being "dark" as it is with our limited perceptions; more about this later). It's rare that star races or star beings live on planets—this is something that is quite restricted to the holographic universe we live in now—planets are normally used for other things. Living on a planet was an Experiment created by the Queen of the Stars—a First Spirit of the Divine Feminine—in order to explore if star beings could live on a beautifully created planet and still be able to travel across the dimensions with thoughts, as all other star beings do. Moreover, living on a planet can sometimes be more challenging than otherwise desired, and the Queen wanted to find out if such beings, whom I call the Namlú'u—the Primordial Spiritual Womankind—could still do this and contain their level of compassion. No death was included in Her Divine Experiment that was taking place in the KHAA. Before that, living on planets was not something star beings did in general—it was a new idea. The human soul group was created specifically for this Experiment, and this soul group was very dear and important to the Queen. She gave these souls freedom to create and a promise to return to the Orion Empire, of which She is the Queen, with no strings attached. This human soul group is of course us—Earth humans.

A fire composite in the KHAA usually splits herself again and thus creates "copies" of herself as she goes along, in order to be able to explore and create even more, but also to protect herself in case one of the soul fragments would be annihilated in a war or by a malevolent being. Here on Earth, we do a similar thing. It is true that each individual here on Earth reincarnates over and over across the lines of time, but each soul of the human soul group is also split into other parts of herself, who also reincarnate again and again and have their own experiences throughout what we call time. The difference between the fire splinters of the Spiritual Universe (the KHAA) and soul splinters in the Physical Universe is that the former always know where another fire composite splinter is and what she does—they are interconnected as One Being. Here on Earth, we have death and amnesia. Therefore, one splinter of oneself is not aware of another, and therefore, each splinter develops differently into different personalities, based on the individual experiences each splinter has. It's done this way with the purpose to create more separation. (Penre, 2019a)

Chapter Five

Near Death Experience [NDE] as Evidence of the Afterlife/Interlife Matrix

The question at the core of Near Death Experiences [NDEs] and Spiritual Dimensions of the Omniverse is—what aspect of the dimensional ecology do NDEs prove?

Are authentic NDEs evidence of a forced reincarnation Birth-Death cycle and Afterlife/Interlife Matrix for Souls imprisoned on planet Earth, as researchers such as Wes Penre would suggest?

Or are authentic NDEs evidence of the nature of the real Spiritual Dimensions that are fully available to humanity after bodily death?

> Beyond books, Near Death Experiences (NDEs) are no stranger to scholarly, peer-reviewed literature. Over 900 articles on NDEs were published in scholarly literature prior to 2005, gracing the pages of such varied journals as *Psychiatry, The Lancet, Critical Care Quarterly, The Journal for Near-Death Studies, American Journal of Psychiatry, British Journal of Psychology, Resuscitation, and Neurology.*
>
> In the 30-year period after Dr. Raymond Moody published *Life after Life,* 55 researchers or teams published at least 65 studies of over 3500 NDEs. (Miller, 2012)

According to the University of Virginia Medical School's Division of Perceptual Studies,

Near-Death Experiences [NDEs] are intensely vivid and often life-transforming experiences, many of which occur under extreme physiological conditions such as trauma, ceasing of brain activity, deep general anesthesia, or cardiac arrest in which no awareness or sensory experiences of any kind should be possible according to the prevailing views in neuroscience.

A near-death experience, or NDE, is a common pattern of events that many people experience when they are experiencing intense threat, are seriously ill, or come close to death. Although NDEs vary from one person to another, they often include such features as the following:

- feeling very comfortable and free of pain,
- a sensation of leaving the body, sometimes being able to see the physical body while floating above it,
- the mind functioning more clearly and more rapidly than us,
- a sensation of being drawn into a tunnel or darkness,
- a brilliant light, sometimes at the end of the tunnel,
- a sense of overwhelming peace, well-being, or absolute, unconditional love,
- a sense of having access to unlimited knowledge,
- a "life review," or recall of important events in the past,
- a preview of future events yet to come, and
- encounters with deceased loved ones, or with other beings that may be identified as religious figures. (University of Virginia Medical School's Division of Perceptual Studies, n.d.)

Can Experiences Near Death Furnish Evidence of Life After Death?

Replicable parapsychology studies using the scientific method have established qualified near death experiences (NDEs) as *prima facie* evidence of an Earthly human Afterlife, or more accurately an Interlife.

In *"Can Experiences Near Death Furnish Evidence of Life After Death?"* (1999-2000), University of Virginia (UVA) Medical School para-psychology founder Dr. Ian Stevenson MD and two UVA colleagues, Emily Williams Cook and Bruce Greyson, published the evidentiary keys linking near-death experiences (NDEs) to formal scientific evidence of an Afterlife/ Interlife.

Dr. Kelly and his colleagues Greysun and Stevenson (1999-2000) write,

> Most people who have a near-death experience (NDE) say that the experience convinced them that they will survive death. People who have not had such an experience, however, might not share this conviction. Although all features of NDEs, when looked at alone, might be explained in ways other than survival, there are features in particular that we believe suggest the possibility of survival, especially when they all occur in the same experience.
>
> These [three] features are: enhanced mental processes at a time when physiological functioning is seriously impaired; the experience of being out of the body and viewing events going on around it as from a position above; and the awareness of remote events not accessible to the person's ordinary senses.
>
> We emphasize however, that near-death experiences can provide only *indirect* evidence of the continuation of consciousness after death: because the persons having these experiences have

lived to report them, they were therefore not dead, however close they might have been to that condition.

Nevertheless, near-death experiences of the type we have described, together with other kinds of experiences suggesting survival after death, provide convergent evidence that warrants our taking seriously the idea that consciousness might survive death. (Kelly, Greysun, & Stevenson, 1999-2000)

Dr. Ian Stevenson and his two colleagues continue their focus in *"Do Any Near-Death Experiences Provide Evidence for the Survival of Human Personality after Death?* [edited]*"* with the case of Al Sullivan and his hospital operating room-based near-death experience (NDE).

Al Sullivan's NDE exhibits key interactive components—(1) enhanced mental activity while bodily disabled (2) Out-of-body, and (3) Paranormal experiences—to near-death experiences (NDEs) that provide evidence of an Afterlife/Interlife in the Spiritual Dimensions

Al Sullivan recounts his NDE:

I began my journey in an upward direction and found myself in a very thick, black, billowy smoke like atmosphere. The smoke seemed to surround me no matter what way I turned, yet it was not going to deter me as far as I was concerned....

As I continued on my journey, I rose to an amphitheater like place. It had a wall directly in front of me to prevent me from going into it. Behind this wall, a very bright light shone.

As I tried to get closer to this wall, I noticed three humanlike figures at my immediate left....

I was able to grasp the wall and look over it into the area the wall was blocking. To my amazement, at the lower left-hand side was, of all things, me. I was laying [sic] on a table covered with light blue sheets and I was cut open so as to expose my chest cavity. It was in this cavity that I was able to see my heart on what appeared to be a small glass table. I was able to see my surgeon, who just moments ago had explained to me what he was going to do during my operation. He appeared to be somewhat perplexed. I thought he was flapping his arms as if trying to fly....

It was at this point I noticed one of the three figures I saw on my arrival to the wall was that of my brother-in-law who had died almost two years before....

It was then that I turned my attention to the lower right-hand side of the place I was at. I saw the most brilliant yellow light coming from, what appeared to be, a very well lit tunnel.... The light that came from the tunnel was of a golden yellow hue and although the brightest I had ever looked into, it was of no discomfort to the eyes at all.

Then, preceded by warmth, joy, and peace and a feeling of being loved, a brown cloaked figure drifted out of the light toward me. As my euphoria rose still more, I, much to my delight, recognized it to be that of my mother.

My mother had died at age thirty-seven when I was seven years old. I am now in my fifties and the first thought that came to my mind was how young my mother appeared. She smiled at me and appeared to be shaping words with her mouth and these was [sic] not audible to me.

Through thought transfer, we were soon able to communicate. All at once my mother's expression changed to that of concern. At this point she left my side and drifted down toward my surgeon. She placed the surgeon's hand on the left side of my heart and then returned to me.

I recall the surgeon making a sweeping motion as if to rid the area of a flying insect. My mother then extended one of her hands to me, but try as I might I could not grasp it. She then smiled and drifted back toward the lit tunnel...

Later it emerged that the surgeon operating on Al Sullivan in fact did "flap" his arms in the form of wings when he use his elbows to give instructions to assistants in the operating room if he had not yet "scrubbed in". (Cook, Greyson, & Stevenson, n.d.)

In "Near-Death Experiences With Reports Of Meeting Deceased People", Dr. Stevenson's colleague, E.W. Kelly notes,

A fourth feature of NDEs is likewise suggestive of survival and provides additional opportunities for comparing competing explanations of NDEs in general. In a substantial number of instances, people have reported seeing, hearing, or sensing the presence of one or more recognized deceased people, usually loved ones, during the NDE.

For example, one person who suffered cardiac complications during routine surgery said : "I was in the brightest place I have ever seen . . . . I actually saw my friend Bill . . . coming towards me with his arms outstretched—he looked so healthy and smiling (not the way he looked before he died)."

Another person—a utility lineman who was electrocuted—described his experience briefly as follows: "I could see myself laying [sic] on the ground, very plainly. It was a feeling as though I was floating in a tunnel with a couple of white doves flying along. The further down the tunnel I went, it became lighter, and when I got closer to the end I could see a human figure, which was my [deceased] mother, motioning for me to go back— not to come any further. After she motioned me back, I came to, with the paramedics working over me."

Similar experiences—of seeming to see, hear, or otherwise be in the presence of a deceased loved one—are not uncommon among waking, apparently healthy persons and among dying patients who actually go on to die shortly after the apparitional experience. Most people who have such an experience during an NDE are convinced that they have been in the presence of a deceased loved one whose consciousness has apparently survived physical death in some form capable of being experienced or perceived by a still-living person.

As one of our participants said: "Seeing my dad made me truly believe that when a person dies a love[d] one will appear to ease the transition to the other side . . . . I knew I was dying and that he had come to get me." (Kelly, 2017)

In "Evidence of the Afterlife: The Science of Near-Death Experiences", Jeffrey Long MD reports

The NDERF [Near Death Experience Research Foundation] survey asked, "Did you meet or see any other beings?" In response, 57.3 percent answered "Yes." When NDErs encounter deceased beings, most are deceased relatives as opposed to friends or loved ones. Some NDErs encounter seemingly familiar beings, but they cannot recall having previously met them. Later in their lives some NDErs recognize a picture of a deceased relative as the being they encountered in their NDE. The relative may have died years or even decades before the NDEr was born. (Long, n.d.)

Kenneth Ring, professor of psychology at the University of Connecticut, ... set out to provide a substantial scientific foundational basis for the NDE, introducing satisfactory sampling procedures and comparison groups, and quantifying variables. He and his staff interviewed more than a hundred people who had come close to death. The aspects they investigated included the incidence of NDEs; whether NDEs are influenced by the circumstances that cause them; and the possible influence of religious beliefs. They also compared NDEs reported during suicide attempts, illness, and accident and went on to investigate the phenomenon of subsequent life changes.

To measure the experience Ring constructed the Weighted Core Experience Index (WCEI), based on the components identified by Moody.61 The stages he incorporated were:

- the sense of being dead,
- feelings of peace and well-being,
- body separation,
- entering the darkness,
- encountering a presence or hearing a voice,
- taking stock of one's life, seeing or being enveloped in the light,
- seeing beautiful colours,
- entering the light, and

- encountering visible "spirits".

Each of the components included in the WCEI were assigned a value. The range of the experience was between 0 and 29, 0 indicating "no components experienced" and 29 signifying a "very deep experience". (Sartori, 2015)

— British medical professor and resuscitation specialist Sam Parnia led the AWARE—AWAreness during REsuscitation study [that] was begun in 2008 and involved fifteen hospitals in the UK, the U.S., and Austria. Two thousand and sixty cardiac arrests were recorded, from which 330 patients survived.

Patients were enrolled between 2008 and 2012. An initial interview was carried out while they were still in the hospital, if possible, from three days to four weeks following the experience; otherwise, they were contacted by telephone, between three and twelve months later. In the first interview, they were asked if they remembered anything from the time of their unconsciousness. Fifty-five people answered yes, all having been verifiably unconscious as measured by lack of response to stimuli, including pain caused by chest compressions.

A second interview was carried out, based on Bruce Greyson's 16-item NDE Scale[6] to ascertain whether they had actually had experienced an NDE. Their memories while unconscious were divided into three categories:

1. detailed non-NDE memories without recall or awareness of events that happened around them while they were unconscious
2. detailed NDE memories without recall or awareness

of events

3. detailed NDE memories with detailed auditory and/ or visual awareness and recall of events.

Forty-six people reported non-NDE memories, which could be divided into seven main themes: fear; animals and plants; a bright light; violence or a feeling of being persecuted; *deja vu* experiences; seeing family members; recalling events that likely occurred after recovery.

Seven reported NDE memories without recall or awareness of events, describing time passing either slower or faster, feelings of peace and pleasantness, being separated from their bodies, having heightened senses, and other sensations typical of NDEs. Parnia *et al* quote one account:

I have come back from the other side of life... God sent (me) back, it was not (my) time—(I) had many things to do... (I traveled) through a tunnel toward a very strong light, which didn't dazzle or hurt (my) eyes... there were other people in the tunnel whom (I) did not recognize. When (I) emerged (I) described a very beautiful crystal city... there was a river that ran through the middle of the city (with) the most crystal clear waters. There were many people, without faces, who were washing in the waters... the people were very beautiful... there was the most beautiful singing... (and I was) moved to tears. (My) next recollection was looking up at a doctor doing chest compressions. (Wehrstein, 2018a)

In Ernesto Bozzano's *Phénomènes Psychiques au Moment de la Mort* (Psychic Phenomena at the Moment of Death), apparitions of the deceased in deathbeds presented cases of deathbed visions, or visions seen by dying persons, and, in some cases, by individuals around the dying person.

The cases were organized in six groups:

1. apparitions of persons known to be dead, seen only by the dying person;
2. apparitions of individuals that no one knew had died, perceived only by the dying person;
3. apparitions perceived both by dying person and other persons at the scene;
4. apparitions related to information communicated via mediums;
5. apparitions only seen by relatives of the dying person; and
6. apparitions seen after the death close to the body

One case in the first group concerned the death of a man called Lloyd Ellis, and included, in addition to the vision, a prediction of time of death. It is cited here from the original source in English used by the author:

Lying in an apparent sleep one night... he woke up suddenly and asked his mother—"Where is my father?" She answered him, tearfully, "Lloyd dear, you know your dear father is dead. He has been dead for more than a year now." "Is he?" he asked, incredulously. "Why! he was in the room just now, and I have an appointment with him, three o'clock next Wednesday." And Lloyd Ellis died at three o'clock on the following Wednesday morning. (Alvarado, 2017)

Deathbed Visions

Early research by William Barrett, a professor of physics at the Royal College of Science in Dublin on deathbed visions was based on the observations his obstetrician-wife had of women who died giving birth; some described visions of deceased relatives that had seemingly arrived to receive them. One such account is the following:

> Suddenly she looked eagerly towards one part of the room, a radiant smile illuminating her whole countenance. "Oh, lovely, lovely," she said. I asked, "What is lovely?" "What I see," she replied in low, intense tones. "What do you see?" "Lovely brightness – wonderful beings." Then – seeming to focus her attention more intently on one place for a moment – she exclaimed, almost with a kind of joyous cry, "Why it is father! Oh, he is so glad I am coming; he is so glad."
>
> Another case that occurred shortly before this woman died:
>
> Her husband was leaning over her and speaking to her, when pushing him aside she said, "Oh, don't hide it; it is so beautiful." Then turning away from him towards me, I being on the other side of the bed, Mrs. B. said, "Oh, why there is Vida," referring to a sister of whose death three weeks previously, she had not been told. They had carefully kept this news away from Mrs. B. owing to her serious illness.
>
> Here is an example from the U.S.:
>
> A female cardiac patient in her fifties knew that she was dying and was in a discouraged, depressed mood. Suddenly, she raised her arms and her eyes opened wide; her face lit up as if she was seeing someone she hadn't seen for a long time. She said,

"Oh, Katie, Katie." The patient had been suddenly roused from a comatose state, she seemed happy, and she died immediately after the hallucination. There were several Katies in this woman's family: a half-sister, an aunt, and a friend. All were dead.

Below is another example of an American, who suffered from a painful, deadly illness. The witness reported the following:

Well, it was an experience of meeting someone whom he deeply loved. He smiled, reached up, and held out his hands. The expression on his face was one of joy. I asked him what he saw. He said his wife was standing right there and waiting for him. It looked as though there was a river and she was on the other side, waiting for him to come across. He became very quiet and peaceful—serenity of a religious kind. He was no longer afraid. (Haraldsson, 2017)

American Singer-Songwriter Pam Reynolds' near-death experience (NDE) during surgery was, as measured by the Greyson NDE scale, particularly deep. The maximum score possible is 32. Reynolds scored 27.

While Reynolds was clinically dead, she was experiencing the following NDE:

Then they [deceased relatives] were feeding me. They were not doing this through my mouth, like with food, but they were nourishing me with something. The only way I know how to put it is something sparkly. Sparkles is the image that I get. I definitely recall the sensation of being nurtured and being fed and being made strong. I know it sounds funny, because obviously it wasn't a physical thing, but inside the experience I felt physically strong, ready for whatever.

My grandmother didn't take me back through the tunnel, or even send me back or ask me to go. She just looked up at me. I expected to go with her, but it was communicated to me that she just didn't think she would do that. My uncle said he would do it. He's the one who took me back through the end of the tunnel. Everything was fine. I did want to go.

But then I got to the end of it and saw the thing, my body. I didn't want to get into it... It looked terrible, like a train wreck. It looked like what it was: dead. I believe it was covered. It scared me and I didn't want to look at it.

It was communicated to me that it was like jumping into a swimming pool. No problem, just jump right into the swimming pool. I didn't want to, but I guess I was late or something because he [the uncle] pushed me. I felt a definite repelling and at the same time a pulling from the body. The body was pulling and the tunnel was pushing... It was like diving into a pool of ice water... It hurt!

The assistants finished the operation to the sound of rock music, Reynolds recalled:

They were playing "Hotel California" and the line was "You can check out anytime you like, but you can never leave." I mentioned [later] to Dr Brown that that was incredibly insensitive and he told me that I needed to sleep more. [laughter] When I regained consciousness, I was still on the respirator. (Wehrstein, 2017b)

Eben Alexander, MD

Eben Alexander MD is a former associate professor at Harvard Medical School who on November 10, 2008 underwent a seven-day coma during which the neocortex of his brain was non-functional.

> After an interdeterminate period, [Alexander] observed an object that turned slowly, radiating filaments of white-gold light, and accompanied by indescribably beautiful music—the "Spinning Melody" as he later named it. He realized he was looking through the light, then began to soar upwards, seeing lush green Earth-like countryside, streams, waterfalls, and joyful people—a place he named the "Gateway".
>
> As he flew, he saw a beautiful girl with him, riding on a delicate surface that was covered with intricate patterns and indescribable colours, like the wing of a butterfly. Millions of butterflies were all around. The Girl on the Butterfly Wing, as he later named her, gave him a message: "You are loved and cherished, dearly, forever. You have nothing to fear. There is nothing you can do wrong".....
>
> He continued to a skyscape of pink-white fluffy clouds, and far above them saw flocks of transparent shimmering orbs flying, which he realized were advanced beings. Sight and hearing were not separate here; he felt he could not sense anything here without becoming part of it. Each time he thought of a question, the answer "came instantly in an explosion of light, color, love, and beauty". He came to an infinite void, dark and yet full of light, which he named the "Core". Here he was given information that, he writes, will take him the rest of his life to unpack: that there is not one universe but many, with many forms of life; that loves lies at the centre of all of

them; and that although evil exists—including on Earth, because there is free will—it is relatively rare.

As he moved down through great walls of clouds he heard countless beings praying for him, helping him keep his spirits up. Heaven, he was promised, would always be with him. He began to see faces, and knew they were people important to him on Earth, though he could not identify them yet. The feeling of freedom was replaced by one of obligation, and the hyper-reality of the experience diminished. "My mind—my real self—was squeezing its way back into the all too tight and limiting suit of physical existence, with its spatiotemporal bounds, its linear thought, and its limitation to verbal communication", he writes.....(Wehrstein, 2018c)

In "Cosmological Implications of Near-Death Experiences", and "Implications of Near-Death Experiences for a Postmaterialist Psychology", UVA researcher Bruce Greyson (2017a) argues that materialist psychology, like classical physics, does not adequately explain phenomena such as a near-death experience [NDE], which might require survival after bodily death to fully account for its evidence.

Bruce Greyson writes,

"Near-death experiences" include phenomena that challenge materialist reductionism, such as enhanced mentation and memory during cerebral impairment, accurate perceptions from a perspective outside the body, and reported visions of deceased persons, including those not previously known to be deceased.

Complex consciousness, including cognition, perception, and memory, under conditions such as cardiac arrest and general anesthesia, when it cannot be associated with normal brain function, requires a revised cosmology anchored not in 19th-century classical physics but rather in 21st-century quantum physics that includes consciousness in its conceptual formulation.

Classical physics, anchored in materialist reductionism, offered adequate descriptions of everyday mechanics but ultimately proved insufficient for describing the mechanics of extremely high speeds or small sizes, and was supplemented a century ago by quantum physics. Materialist psychology, modeled on the reductionism of classical physics, likewise, offered adequate descriptions of everyday mental functioning but ultimately proved insufficient for describing mentation under extreme conditions, such as the continuation of mental function when the brain is inactive or impaired, such as occurs near death. (Greyson, 2017a)

Researcher Greyson (2017a) continues,

There is one particular kind of vision of the deceased that calls into question even more directly their dismissal as subjective hallucinations: cases in which the dying person apparently sees, and often expresses surprise at seeing, a person whom he or she thought was living, who had in fact recently died. Reports of such cases were published in the 19th century and have continued to be reported in recent years.

In one recent case, a 9-year-old boy with meningitis, upon awakening from a 36-hour coma, told his parents he had been with his deceased grandfather, aunt, and uncle, and also with his 19-yearold sister who was, as far as his family knew, alive and well at college 500 miles away.

Later that day, his parents received news from the college

that their daughter had died in an automobile accident early that morning. Because in these cases the experiencers had no knowledge of the death of the recently deceased person, the vision cannot plausibly be attributed to the experiencer's expectations.

In sum, the challenge of NDEs to the mind-brain production theory lies in asking how complex consciousness, including mentation, sensory perception, and memory can occur under conditions in which current neurophysiological models deem it impossible. This conflict between a materialist model of brain producing mind and the occurrence of NDEs under conditions of general anesthesia and/or cardiac arrest is profound and inescapable.

Only when we expand models of the mind to accommodate extraordinary experiences such as NDEs will we progress in our understanding of consciousness and its relation to the brain. The predominant contemporary models of consciousness are based on principles of classical physics that were shown to be incomplete in the early decades of the 20th century.

However, the development of post-classical physics over the past century offers empirical support for a new scientific conceptualization of the interface between mind and brain compatible with a cosmology in which consciousness is a fundamental element. (Greyson, 2017b)

Chapter Six

Out of Body Experiences (OOBEs), Instrumental Transcommunication (ITC), Chronovision: Evidentiary Proof of Wes Penre's Hypothesis about the false Afterlife/Interlife Reincarnation Matrix

Out of Body Experiences (OOBEs)

"When the relay station and filter, which is our brain, stops functioning and the body is returned to its individual atoms, our conscious and subconscious mind become our new external reality" (Ziewe, n.d.a)

Out-of-body-experiencer Jurgen Ziewe writes,

> I communicate a lot with dead people during OOBEs. It has become the main source of information for me. Often, when the opportunity of an OOBE presents itself, I see myself as a reporter interviewing people in much the same way as a journalist would interview people attending a certain event. This is a crucial aspect of obtaining information about the next dimensional realities. Not only do you get information via the interviewees' experience, but you get the complete picture. By talking to them, which is mostly done via thought anyway, you can actually watch their background played out in front of you like a movie without having to ask many questions.
>
> I also found that the people I talked to rarely showed objec-

> tion to being approached, and on occasion, I even established quite intensive bonds in record time, simply by relating to them on quite personal levels, getting their "picture" and being able to help them. Some of the people I encounter are not even dead, they are simply asleep, but I still deal with them and can help. (Ziewe, (n.d.b)

Researcher Karen Wehrstein writes, *In Journeys Out of the Body*, [Robert] Monroe categorized his destinations into three "locales" and calculated the percentage of his visits spent in each:

> Locale I (31%): the material world, the world represented to us by our physical senses, containing no strange beings, events, or places. All veridical information is gathered here as it can be confirmed by others in the material world.
>
> Locale II (59.5%): a non-material environment with unknown bounds, which contains all aspects of what are known as heaven and hell, apparent physical spaces and objects, and entities with whom communication is possible. Measurement of time or space is irrelevant, and thought, however much influenced by unrepressed emotion, instantly creates one's reality. In OBEs, one tends to be drawn to this locale.
>
> Locale III (8.9%): an apparently parallel material world, different from ours in technology and social customs.
>
> Monroe later simplified this schema to two categories:
>
> 1. "local traffic" or places nearby (that is, in the material world), and
> 2. "interstate" or highway, meaning places more distant and immaterial.
>
> Monroe did not refer either to an immortal soul or to rein-

carnation, but was certain of survival beyond physical death: in the OBE state, he met deceased persons and revisited apparent previous lives of his own. His cosmology includes heaven or hell only as part of what he called "the belief system territories", in which strong believers receive rewards or penalties according to their beliefs.

He believed that it was possible for human minds to refuse to see or accept that they have left a body, and therefore be "unable to free themselves from the ties of the Earth life system," remaining stuck in limbo until they realize their situation. (Wehrstein, 2018b)

OOBEs as Evidence for an Afterlife in the Spiritual Dimensions

Because Out of Body Experiences (OOBEs) can be free-will visits to the Spiritual Dimensions, OOBEs are a different category of scientific evidence for the existence of the afterlife than near-death experiences [NDEs].

First-person accounts of visitors' explorations to the human afterlife during OOBEs can provide authentic field notes and research as to what the nature and reality of the human afterlife is like for the human souls that inhabit it.

Parapsychology research shows that scientifically-derived information about the immediate human afterlife is currently systematically suppressed by those institutions charged by modern society with preparing humans for the afterlife: organized religions, universities, scientific and medical organizations, and institutionalized atheism and agnosticism of modern society on Earth.

Suppression of the scientific reality of the life of the soul in the Omniverse arose for complex exopolitical and historical reasons and endures on Earth largely for self-serving institutional reasons of the vested interests and is leading to systematic, untold, and continuing misery for the souls of deceased humans in the immediate human afterlife.

An Immediate Afterlife Where Souls are not Aware That Their Bodies Have Died

In *"Vistas of Infinity—How to Enjoy Life When You Are Dead"*, Out-of-body-experiencer Jurgen Ziewe (n.d.a) states that he conducts OOBE visits to an immediate human Afterlife zone in the Spiritual Dimensions where Souls are not aware their bodies have died.

Ziewe writes,

> By the time you have finished reading this introduction three thousand people will have died on our planet. By the end of the day their numbers could populate a small town if they were to gather in one place. The vast majority of these people will have no idea of what has happened to them and what is going to happen after their body has been recycled. Our ignorance about our future destiny when we die is problematic because it is the source of much fear, uncertainty and suffering. (Ziewe, n.d.a)
>
> Ziewe (n.d.a) summarizes:
>
> Misconceptions about Life after Death—Afterlife states are as numerous as human beings, and as complex. The most important things we will have to reassess, and most probably shake off, is a widely held belief that our afterlife will magically change everything and reset all our parameters. Individual reports by near-death-experiencers and channelling mediums are very specific and will almost certainly never apply to ourselves. Some of the medium reports I

found were very confirmative to what I have experienced. There is still the rather romantic view that with the end of our life our suffering will come to an end and we will live in eternal peace.

RIP is a fallacy that has been implanted in us from early childhood, mainly as a result of ignorance and our inability to look beyond the great curtain. When people proclaim at funeral services, “Well, at least now he/she is at peace”, mostly this is not the case unless the deceased was already at peace at the time of death. What in fact happens is that nothing changes and at the same time everything changes ... in some ways....

On the whole, there are few giveaway signs that they are no longer alive in a physical world. If told so, they may be adamant in their denial by pointing out that everything around them is as real as it could be and they are fully awake and aware and that there is no way that they could be dead because the evidence shows otherwise. (Ziewe, n.d.a)

Scientist Emmanuel Swedenborg [1688-1772] uses his OOBEs into the dimensional ecology of the Spiritual Dimensions to confirm that in the immediate Afterlife there is a zone where souls whose bodies have died still believe they are alive on Earth.

In *Heaven and Hell* Swedenborg writes,

I have talked with some people on the third day after their death,... I talked with three whom I had known in the world and told them that their funeral services were now being planned so that their bodies could be buried. When they heard me say it was so that they could be buried, they were struck with a kind of bewilderment. They said that they were alive, and that people were burying what

> had been useful to them in the world. Later on, they were utterly amazed at the fact that while they had been living in their bodies they had not believed in this kind of life after death, and particularly that this was the case for almost everyone in the church.
>
> Some people during their earthly lives have not believed in any life of the soul after the life of the body. When they discover that they are alive, they are profoundly embarrassed. However, people who have convinced themselves of this join up with others of like mind and move away from people who had lived in faith. (Swedenborg, n.d.)

Swedenborg's OOBEs provide him with a topology of the Afterlife. The Swedenborg Foundation summarizes this mapping of Heaven as follows:

> Swedenborg calls the realm we enter immediately after death the world of spirits, an intermediate realm situated between heaven and hell. It can be thought of as a "sorting out" zone from which spirits go to either heaven or hell. He describes three states that people might pass through in this realm.
>
> In the first state, people are essentially the same as they were in life. They have all of their memories, they have the same beliefs and attitudes toward things, and they may even manifest the same surroundings that they had on Earth. Swedenborg says that this is why some people who have died aren't even aware that they are in the spiritual world and may try to deny it if they are told so by an angel.
>
> When people first enter the spiritual world, they often meet friends or relatives who crossed over before them. Spouses will

be reunited, although not necessarily forever. The spiritual world is a place where a person's inner nature becomes the whole of their being. If two people were truly of one mind on Earth, they will live together as spouses in heaven too. However, if they were not happily married, or if their personalities are fundamentally different, they will eventually part ways. Those who did not find love on Earth, Swedenborg says, will eventually find their perfect match in heaven—no one is ever alone unless they wish to be.

Friends and relatives become the new arrival's guide to the spiritual world, and, with the help of good spirits, the person's true inner nature will gradually be revealed. This first state might last anywhere from a few hours to a year or more, depending on how long it takes for a person's outer nature (what they outwardly say and do) to harmonize with their inner nature (what they truly feel and believe). Anyone who has become totally transparent in this life, whether transparently loving or transparently hateful, is fully ready for either heaven or hell, and goes straight in.

In the second state after death, the person becomes aware of the deeper parts of his or her inner nature. They start saying what they really think and act according to what they feel without worrying about appearances or making other people happy. (Swedenborg Foundation, n.d.)

Swedenborg's bipolar destination topology of the human Afterlife as "Heaven & Hell" has been questioned by more modern OOBErs such as Robert Monroe who, unlike Swedenborg, are not filtering their observations through the matrix of the Old and New Testament Bible and its emphasis on heaven and hell.

Robert Monroe uses OOBEs to document Afterlife levels:

when a human dies and leaves the physical body permanently, he becomes consciously aware of his energy body. Whatever his belief system during his last life on Earth, he will be attracted to the energy level which resonates with his own belief system at the moment of transition from the physical to the non-physical.

If a person is strongly catholic, for example, she would land in a realm where there is a catholic after life reality created by like-minded souls.

On one journey—triggered by his own sexual arousal—Robert Monroe visited a lower vibrating level, where bodies upon bodies were engaged in constant intercourse – like group sex, obsessed with sex—never getting enough satisfaction.

Then he describes souls who have died, but have not realized or who cannot accept their physical death. They seem to be stuck in the near Earth vibrations and are unable to see the light that would lead them into higher realms. Monroe himself became a helper to these so-called Earthbound souls, trying to make them realize that they have survived physical death.

Even in the after world, it seems, helpers are needed to give proof of life after death to the deceased non-believers of an after-life. No matter where a recently passed human soul would land, there would always be helpers from higher levels gracefully and patiently trying to make them see another perspective of their current awareness of existence. Always focused on the intent to raise their vibration. (Lucid-Mind-Center.com, n.d.)

Afterlife researcher and retired trial lawyer Victor Zammit who investigates OOBEs as well as over 20 other sources of replicable scientific information about the human afterlife concludes in *"What happens when we die"* in *"A Lawyer Presents the Case for the Afterlife Irrefutable Objective Evidence":*

> There are different levels or "spheres" in the afterlife – from the lowest vibrations to the highest. On physical death we go to the sphere which can accommodate the vibrations we accumulated throughout our life on Earth. Simplistically put, most ordinary people are likely to go to the "third" sphere – some people call it the "Summerland." The higher the vibrations, the better the conditions – this will take us to the higher spheres. We are informed that the higher spheres are too beautiful to even imagine. For those with very, very low vibrations, very serious problems do exist.
>
> Hell for eternity and eternal damnation were invented by men to manipulate the hearts and the minds of the unaware – they do NOT exist. Whilst there ARE lower spheres in the afterlife that are particularly dark, unpleasant, and even horrific – some call them "hell" – ending down there is NOT for eternity. There is always help available for any soul willing to learn the lessons of kindness and unselfishness.
>
> Soon after crossing over, you will experience a life-review. In your life review, you will experience all of your thoughts, words, and deeds and effects they had on others. Nobody judges you. You judge yourself by comparing the reality of your life and the effects it had on others with what you set out to do.
>
> You also will have the opportunity to go to the Halls of Learning, and continue to do spiritual work – helping those crossing over

> or helping others less informed. You may like to do rescue work – informing those lost in the darker realms and who qualify to be in the sphere of the light to come up towards the light. You can be creative in how you spend your time.
>
> Those who consistently abused and harassed others will have to face their victims in the afterlife to ask for forgiveness. After the severest retribution, the transgressors will have to apologize and seek forgiveness by the victims before they are allowed to make any progress. (Zammit, n.d.)

Robert Monroe's own OOBE reports from the human Afterlife in his book *Far Journeys* are congruent with the essence of Afterlife researcher Robert Zammit's conclusions that the reality of an individual's Afterlife is a creation of their own inner soul life and values.

> On his extensive journeys into other worlds, Monroe meets a guide whom he calls INSPEC friend. The INSPEC helps him learn Non Verbal Communication (NVC). Monroe relives significant situations of his previous lives. At another occasion, he lives through a real life situation over and over again until he gets his motivation and thus his resulting action right. A bit like the movie *Groundhog Day*.
>
> Very frequently on his astral travels, Monroe meets BB, an entity from the energy system KT-95. BB is kind of hanging around looking for his best friend – AA – who has incarnated many human lifetimes ago. BB tries to get AA back to KT-95, but AA is so addicted to the human life form, that he cannot even remember ever being anything else but a human before.
>
> Monroe takes it upon himself to teach BB what it means to

be a human. He shows BB the sleeping schools, where the humans go to during the sleeping hours to learn how to manage a situation in their daily lives.

He shows him the different stages of the afterlife and introduces BB to Lou, who is a highly developed soul and to Charlie who knows that he is dead but who is only waiting to get back to a physical life.

BB explains to Monroe, that AA and he came on a travel tour of the energy system Earth – like some space tourists. As part of their travel information, they received a creation story of the Earth.

That creation story basically says that the humans have been created by aliens to produce emotional energy which serves as fuel for the aliens' energy system. (I wonder where the idea for the movie MATRIX came from.)

In the last few chapters, he has a glimpse or a journey into the possible future of the Earth in year 3000. While there, Monroe can take on the form of any body—animal, wind, water, etc.—of the Earth for learning purposes. Food is not necessary and can be made for pure pleasure from dirt at will. The human bodies go into a kind of hibernation while the mind/soul is busy exploring other aspects of reality. Fascinating! BB is there and his wife Nancy and also AA, but there is some kind of barrier between Monroe and AA.

As he "travels" back to Earth to the here and now, he observes in the outer rings many entities that have come to observe this unique universal event in Earth history. Monroe calls it "The Gathering", which apparently only happens every 87,000,000 years. (Lucid-Mind-Center.com, n.d.)

Instrumental Transcommunication [ITC] and Chronovision: How the Afterlife/Interlife Matrix and the Time Matrix Intersect

Afterlife researchers Victor and Wendy Zammit have documented the technological forensic evidence of the Afterlife/Interlife Matrix provided by Instrumental Transcommunication [ITC], as has researcher Ervin Lazlo. Prof. Ernst Senkowski notes that Tesla, Edison, and Marconi all pursued ITC as a research tool for non-local communication "with other worlds".

Victor and Wendy Zammit write,

> According to Mark Macy, a leading researcher in this area, throughout the 1990s, the research laboratories in Europe received extended, two-way communication with spirit colleagues. This was almost daily through telephone answering machines, radios, and computer printouts. His book *Miracles in the Storm* outlines how scientists working for the International Network for Instrumental Transcommunication (INIT), which he founded in 1995, received communication from the afterlife:
>
> > Pictures of people and places in the afterlife on television that either appeared clearly on the screen and remained for at least several frames, or which built up steadily into a reasonably clear picture over multiple frames; text and picture files from people in the afterlife which appeared in computer memory or were planted on disk or similar recordable media; and text and images of people and places in the afterlife by way of fax messages. (Zammit, n.d.)

Instrumental Transcommunication ITC as Nonlocal Communication

Researcher Ervin Laszlo focuses on Instrumental Transcommunication ITC as a modality for "non-local" communication and writes,

> The authenticity of ITC is not entirely beyond doubt, but the evidence for it is sufficiently robust to merit sustained investigation. In this writer's view, ITC might be a hitherto unexplored domain of nonlocality, a form of nonlocal communication.
>
> To find a scientific basis for the ITC phenomenon, it must be connected with theories in physics. Transcommunication cannot be connected with classical physics, given that the latter is based on a paradigm that views phenomena not directly traceable to sensory experience as highly suspect, if not clearly illusory. But at the cutting edge of contemporary physics, many things and processes are acknowledged as elements of reality, even when they are intrinsically unobservable. Most pertinently, theories in particle as well as cosmological physics make reference to a field or dimension that subtends the world of the quantum, hitherto considered the lowest level of physical reality. This field or dimensions, variously termed "physical space-time," "nuether," "hyperspace," or "atemporal space" may be responsible for the phenomena of nonlocality in the microdomain of the quantum, as well as in the mesodomain of life and the macrodomain of the universe. (Lazlo, n.d.)

Tesla, Marconi, Edison, and ITC

Professor at the Technical University (Fachhochschule) Bingen, Ernst Senkowski, Ph.D., of Mainz, Germany, defines ITC as "Technically supported contacts with 'beings, entities, Informational Structures' that are normally not accessible" (Senkowski, n.d.).

Dr. Senkowski writes,

> Telecommunication devices (audio recorders, radio receivers, telephones, video recorders, TV receivers, and computers) behave irregular[ly in] delivering voices, images, and texts of different, sometimes excellent quality coming from nowhere.
>
> The history of ITC: Tesla, Marconi, Edison suspected [it might be possible] to contact other worlds by electromagnetic means. Psychic mediums predicted the future development of suitable apparatus. Unexplained signals have been observed since WW I. Partially successful tests were realized in the 20s, 30s, 40s. First voices on tapes were documented independently during the 50s in Italy, USA, Sweden. TV-video images and computer texts appeared later. Official research is not known. Attempts to realize voices on tapes are carried out in at least 12 countries by single operators, small groups, and some bigger associations. Around 60 monographs have been published in seven languages. Only very few operators are scientifically interested. About 10 people have so far realized excellent results.
>
> There are three open questions:
>
> 1. How are ITC events realized? Excellent results of these extraordinary psychophysical, mind-matter, man-machine, or consciousness-reality interactions apparently depend upon psychic faculties of the operators. A simple model using the parapsychological terms ESP and PK describes ITC as a largely unconscious two-step process: a trans-entity from beyond our space-time system establishes a telepathic contact with a terrestrial oper-

ator who psychokinetically injects the trans-information into the electronic device. It cannot basically be excluded that our technical systems are directly manipulated from the "other side" using techniques unknown to us. Understanding might result from: consciousness research, neurosciences, psychophysical theories, quantum theoretical models (Bohm, Jahn / Dunne-Pear / Princeton University); sub-quantum or vacuum physics; 12-dimenstional field theory (Burkhard Heim); theory of morphic fields (Sheldrake); system theories; OBE and NDE; thanatology; "far out" speculations [such as] scalar fields and tachyons.

2. What are the contents of ITC? It seems that some "No Body" wants to convince humanity about Life after Death in other states of consciousness and of the existence of nonhuman beings that could be involved to an uncontrollable extent. Generally [speaking], the contents confirm the fundamentals of mediumistic channeling in technical forms.
3. Who generates ITC? It seems unjustified to "explain" ITC phenomena as a mere result of the hypothetical "unconscious" of the operator or by super telepathy. Trans-entities appear autonomous and [appear] to be consciously alive. The selectivity of trans-information in cases of drop-in communicators cannot be explained and points to that direction. It is probable that the operator as an active transducer unconsciously adds a certain amount of his own associations to the primary message.

Important issues:

- demonstration of yet undeveloped faculties of the human mind;
- multidimensionality of humans;
- holistic interconnectedness and non-separability of All-That-Is--Mental awakening is defined as opening windows to hitherto unknown realities or "parallel worlds," including "past" existences; and
- an ongoing and future metamorphosis or transformation of humanity. (Senkowski, n.d.)

Ernetti, Chronovision, and Evidentiary proof of Wes Penre's hypothesis about the Afterlife/Interlife Matrix

As I noted in my earlier book on the Omniverse, it is highly meaningful for the dimensional ecology of the Omniverse hypothesis that Instrumental Transcommunication (ITCs) electronic voice phenomena (EVP), used for exploring the interlife in the spiritual dimensions, and chronovision, a time-travel technology for exploring timelines in universes of the exopolitical dimensions, intersect in the work of two Italian Catholic priests, Fathers Pellegrino Ernetti and Augustino Gemelli. (Webre, 2015a)

Evidentiary Conclusion: This evidentiary identity of (1) the Afterlife/Interlife, and (2) the Time-Space Hologram, suggests that what the mainstream Earth religions as well as the Afterlife/Interlife researchers have been terming "the Afterlife" is in fact a segmented dimension in the Earth time-space hologram also termed the Afterlife/Interlife Matrix, located in the Saturn-Moon matrix as many scholars such as Immanuel Swedenborg and others have suggested.

Thus, it would appear that Wes Penre's hypothesis that the Afterlife/Interlife Matrix is, in fact, an artificial construct inside our time-space Universe hologram is correct. This hypothesis has been proven by the fact that the Chronovisor can access both (1) events in the future and in the past in our Earth's time-space hologram, as well as (2) events in the Afterlife/Interlife Matrix.

Moreover, as we noted,

"Ernetti himself states how the chronovisor can tune into individual voices and images along a timeline in the time-space hologram of the Exopolitical dimension."

Chronovisor researcher Maddaloni writes:

> With the use of suitable equipment, [which] our team was the first in the world to build. The equipment consists of a series of antennas to allow tuning of the individual voices and images. We know that every human being, from when he was born to when he dies, leaves behind himself a double wake: a visual and a sound, a sort of identity card for each different person. And based on this different identity card one can reconstruct the single person in all his deeds and his sayings, and for this reason we are now able to hear and see again the greatest people in history. (Maddaloni, 1972)

Chapter Seven

Reincarnation and Soul Development in the Afterlife/Interlife Matrix of the Omniverse

Our Universe—and universes in general—function as machines for Soul development, among other universe design criteria (Webre, 2015a).

Incarnation and re-incarnation to experience moral challenges through lives as humans is one standard method by which the development of human souls is achieved in the virtual universes of the multiverse.

To be sure, human beings are not the only exophenotype capable of serving as vehicles for the incarnation of souls.

Divine human souls that originate in the Spiritual dimensions are created as holographic fragments of Source.

Generically, souls are the multidimensional embodiment of a Spirit-consciousness that can "incarnate" or embed itself in a physical exophenotype, be it human, or other higher intelligent exophenotype such as avian, feline, reptilian, Grey and other higher intelligent exophenotypes.

Second density or First density animal, vegetable, and mineral species have souls as well that can survive bodily death and "reincarnate". Think of the souls of cats, trees, mountains, and planets.

Earth Human Soul Reincarnation

According to researcher Karen Wehrstein,

> Contrary to widespread belief, in most reincarnation cases

> the subject is reborn in the same country, and often less than 25 kilometers from the location of the previous existence. Dr. Ian Stevenson has published only four international cases in the collection of 1,700 cases held at the University of Virginia's Division of Personality Studies in which the past-life identity was confirmed, although there might be others that have not been published.
>
> Reincarnation researcher Dr. James G Matlock points out that this rarity suggests that souls only travel long distances between lives for specific reasons. His analysis of fourteen international cases (in which the past-life identity was confirmed) identified four motives, each of which played a part in more than one case (this finding is tentative, as the sample is small):
>
> - to return to be with their previous family, friend or compatriot (four cases),
> - to return to their homeland (four cases),
> - to spread the word about Buddhism (three cases), and
> - to leave homeland (two cases).
>
> The first motive accords well with the observed general tendency for people to reincarnate with people they know. (Wehrstein, 2017a)

The more reasonable question is not *if* human consciousness survives into an Afterlife. The preponderance of available evidence and reported experience is heavily on the side of survival after bodily death. Rather, we should be asking *how* human consciousness survives into an Afterlife, and *how* and *toward what end* the dimensional ecology of the human Afterlife operates.

Former U.S. chrononaut—time traveler—and Mars Astronaut Andrew D. Basiago, JD, MPhil (Cantab), states,

> I know that the Afterlife and reincarnation is not just theoretical and is real! I can remember many of my deaths. I have met 11 "deceased" people who are in the Afterlife, including my father Raymond F. Basiago. As part of my training in DARPA-CIA's Project Pegasus secret U.S. government time travel program, I was regressed to a prior life as a geometry professor at a New England college—possibly Yale, Amherst, or Boudoin—where I experienced firsthand that the soul cannot die. The woman scientist who regressed me stated "DARPA had proved that the soul does not die; we are regressing you to a prior life so in case you die during time travel, you understand your soul will not die." There are differing Afterlife conditions whereby a soul might be Earth-bound after one death in order to reincarnate, and might return to the Godhead after another death. Typically individuals like you and I may have up to 10,000 reincarnations on Earth before we are ready to ascend. (Basiago & Webre, 2019)

Our universe is designed as a machine for the development of souls. A principal mechanism for soul development in our universe and in our Earth time space hologram in particular is through the reincarnation technology. Reincarnation is designed as a complex virtual experience whereby individuated holographic fragments of Source created as souls are embedded in human [and other] avatars on Earth and other planets in our universe.

The Stages of Soul Development

Conventional perspective in Afterlife studies is generally structured exactly backwards to the spiritual reality that our soul is living. As living human beings reading this book, for example, we are spiritual souls having a human experience, and not the opposite—human bodies having spiritual experiences.

Our human souls are in fact holographic fragments of Source that are created as individuated souls who populate the dimensional ecology of the Spiritual Dimensions and the Universes of time, energy, space, and matter where souls take on innumerable interdimensional creative and experiential tasks.

Once created, souls are multi-tasking in the Spiritual Dimensions, and are engaged in the great divine enterprise of Creation and maintenance of the Universes—ranging from the creation of individual galaxies and solar systems, the creation and maintenance of exophenotypes—living species—in the universes, other fauna and flora; and the maintenance of dimensions such as time and space in specific universes.

In my public talks, I have often used the metaphor of a laundry washing machine to describe the effects experienced during successive cycles by a human soul in the Reincarnation machine architected by its successive cycles in the Spiritual dimensions, Astral Dimensions, and Earth Human Birth-Death Cycles.

Earth is a venue for the human soul-spirit birth and death cycle in our holographic time-space universe. According to statistical norms, there are on average about 75-80 million births each year, and 56 million deaths each year, or approximately 153,424 deaths per day on Earth or 1.80 deaths per second on an average day. (World Death Clock, n.d.)

University of Virginia Medical School: Dr. Ian Stevenson, MD Reincarnation Research

Canadian Dr. Ian Stevenson, MD applied scientific method analysis to field studies of cases suggestive of reincarnation in the Afterlife/Interlife Matrix.

> The most frequently occurring event or common denominator relating to rebirth is probably that of a child remembering a past life. Children usually begin to talk about their memories between the ages of two and four. Such infantile memories gradually dwindle when the child is between four and seven years old. There are, of course, always some exceptions, such as a child continuing to remember its previous life but not speaking about it for various reasons.
>
> Most of the children talk about their previous identity with great intensity and feeling. Often they cannot decide for themselves which world is real and which one is not. They often experience a kind of double existence, where at times one life is more prominent, and at times the other life takes over. This is why they usually speak of their past life in the present tense, saying things like, "I have a husband and two children who live in Jaipur." Almost all of them are able to tell us about the events leading up to their death.
>
> Such children tend to consider their previous parents to be their real parents rather than their present ones, and usually express a wish to return to them. When the previous family has been found and details about the person in that past life have come to light, then the origin of the fifth common denominator—the conspicuous or unusual behavior of the child—is becoming obvious.

> For instance, if the child is born in India to a very low-class family and was a member of a higher caste in its previous life, it may feel uncomfortable in its new family. The child may ask to be served or waited on hand and foot and may refuse to wear cheap clothes. Stevenson gives us several examples of these unusual behavior patterns. (Stevenson, n.d.)

Physical traits such as birth marks, according to Dr. Ian Stevenson's research on reincarnation in the Afterlife/Interlife Matrix, might be stigmata or signs of significant events such as wounds that were received by an incarnate Soul during a previous life in the Birth Death cycle.

Dr. Ian Stevenson writes,

> Almost nothing is known about why pigmented birthmarks (moles or nevi) occur in particular locations of the skin. The causes of most birth defects are also unknown. About 35 percent of children who claim to remember previous lives have birthmarks and/or birth defects that they (or adult informants) attribute to wounds on a person whose life the child remembers. The cases of 210 such children have been investigated. The birthmarks were usually areas of hairless, puckered skin; some were areas of little or no pigmentation (hypopigmented macules); others were areas of increased pigmentation (hyperpigmented nevi). The birth defects were nearly always of rare types. In cases in which a deceased person was identified, the details of whose life unmistakably matched the child's statements, a close correspondence was nearly always found between the birthmarks and/or birth defects on the child and the wounds on the deceased person. In 43 of 49 cases in which a medical document (usually a postmortem report) was obtained, it

confirmed the correspondence between wounds and birthmarks (or birth defects). There is little evidence that parents and other informants imposed a false identity on the child in order to explain the child's birthmark or birth defect. Some paranormal process seems required to account for at least some of the details of these cases, including the birthmarks and birth defects.

Because most (but not all) of these cases develop among persons who believe in reincarnation, we should expect that the informants for the cases would interpret them as examples according with their belief; and they usually do. It is necessary, however, for scientists to think of alternative explanations.

The most obvious explanation of these cases attributes the birthmark or birth defect on the child to chance, and the reports of the child's statements and unusual behavior then become a parental fiction intended to account for the birthmark (or birth defect) in terms of the culturally accepted belief in reincarnation. There are, however, important objections to this explanation.

First, the parents (and other adults concerned in a case) have no need to invent and narrate details of a previous life in order to explain their child's lesion. Believing in reincarnation, as most of them do, they are nearly always content to attribute the lesion to some event of a previous life without searching for a particular life with matching details.

Second, the lives of the deceased persons figuring in the cases were of uneven quality both as to social status and commendable conduct. A few of them provided models of heroism or some other enviable quality; but many of them lived in poverty or were otherwise unexemplary. Few parents would impose identification with such persons on their children.

Third, although in most cases the two families concerned were acquainted (or even related), I am confident that in at least 13 cases (among 210 carefully examined with regard to this matter) the two families concerned had never even heard about each other before the case developed. The subject's family in these cases can have had no information with which to build up an imaginary previous life that, it later turned out, closely matched a real one. In another 12 cases the child's parents had heard about the death of the person concerned, but had no knowledge of the wounds on that person. . . .

Fourth, I think I have shown that chance is an improbable interpretation for the correspondences in location between two or more birthmarks on the subject of a case and wounds on a deceased person.

Persons who reject the explanation of chance combined with a secondarily confected history might consider other interpretations that include paranormal processes, but fall short of proposing a life after death. One of these supposes that the birthmark or birth defect occurs by chance and the subject then by telepathy learns about a deceased person who had a similar lesion and develops identification with that person. The children subjects of these cases, however, never show paranormal powers of the magnitude required to explain the apparent memories in contexts outside of their seeming memories.

Another explanation, which would leave less to chance in

the production of the child's lesion, attributes it to a maternal impression on the part of the child's mother. According to this idea, a pregnant woman, having knowledge of the deceased person's wounds, might influence a gestating embryo and fetus so that its form corresponded to the wounds on the deceased person. The idea of maternal impressions, popular in preceding centuries and up to the first decades of this one, has fallen into disrepute. Until my own recent article (Stevenson, 1992) there had been no review of series of cases since 1890 . . . Nevertheless, some of the published cases—old and new—show a remarkable correspondence between an unusual stimulus in the mind of a pregnant woman and an unusual birthmark or birth defect in her later-born child. Also, in an analysis of 113 published cases I found that the stimulus occurred to the mother in the first trimester in 80 cases (Stevenson, 1992). The first trimester is well known to be the one of greatest sensitivity of the embryo/fetus to recognized teratogens, such as thalidomide . . . . Applied to the present cases, however, the theory of maternal impression has obstacles as great as the normal explanation appears to have. First, in the 25 cases mentioned above, the subject's mother, although she might have heard of the death of the concerned deceased person, had no knowledge of that person's wounds. Second, this interpretation supposes that the mother not only modified the body of her unborn child with her thoughts, but also after the child's birth influenced it to make statements and show behavior that it otherwise would not have done. No motive for such conduct can be discerned in most of the mothers (or fathers) of these subjects. (Stevenson, n.d.)

Reincarnation and the Nature of Soul

Dr. Michael Newton has pioneered in research of the Spiritual Dimensions and the nature of the human soul, using access to Soul memories of the Afterlife and Interlife through replicable hypnotic regression of over 7000 subjects. Here are some significant samples of Dr. Newton's research on the nature of Soul.

> We cannot define the soul in a physical way because to do so would establish limits on something that seems to have none. I see the soul as intelligent light energy. This energy appears to function as vibrational waves similar to electromagnetic force but without the limitations of charged particles of matter. Soul energy does not appear to be uniform. Like a fingerprint, each soul has a unique identity in its formation, composition, and vibrational distribution. I am able to discern soul properties of development by color tones, yet none of this defines what the soul is as an entity. (Newton, 2009)
>
> *Newton*: As an Incubator Mother, when do you first see the new souls?
>
> *Soul:* We are in the delivery suite, which is a part of the nursery, at one end of the emporium. The newly arrived ones are conveyed as small masses of white energy encased in a gold sac. They move slowly in a majestic, orchestrated line of progression toward us.
>
> *Dr. Newton:* From where?
>
> *Soul:* At our end of the emporium under an archway the entire wall is filled with a molten mass of high-intensity energy and . . . vitality. It feels as if it's energized by an amazing love force rather than a discernible heat source. The mass

pulsates and undulates in a beautiful flowing motion. Its color is like that on the inside of your eyelids if you were to look through closed eyes at the sun on a bright day.

Dr. Newton: And from out of this mass you see souls emerge?

Soul: From the mass a swelling begins, never exactly from the same site twice. The swelling increases and pushes outward, becoming a formless bulge. The separation is a wondrous moment. A new soul is born. It's totally alive with an energy and distinctness of its own. (Newton, 2009)

Another one of my [high] level V [souls] made this statement about incubation:

I see an egg-shaped mass with energy flowing out and back in. When it expands, new soul energy fragments are spawned. When the bulge contracts, I think it pulls back those souls which were not successfully spawned. For some reason these fragments could not make it on to the next step of individuality. (Newton, 2009)

Once through the tunnel, our souls have passed the initial gateway of their journey into the spirit world. Most now fully realize they are not really dead, but have simply left the encumbrance of an Earth body which has died. With this awareness comes acceptance in varying degrees depending upon the soul. Some subjects look at these surroundings with continued amazement while others are more matter-of-fact in reporting to me what they see. Much depends upon their respective maturity and recent life experiences. The most common type of reaction I hear is a relieved sigh followed by something on the order of, "Oh, wonderful, I'm home in this beautiful place again." There are those highly developed souls who move so fast out of their bodies that much of what I am describing here is a blur as they home into their spiritual destinations. These are the pros and, in my opinion, they are a distinct minority on Earth. The average soul does not move that rapidly and some are very hesitant. If we exclude the rare cases of highly disturbed spirits who fight to stay connected with their dead bodies, I find it is the younger souls with fewer pasts.

Most of my subjects report that as they emerge from the mouth of the tunnel, things are still unclear for awhile [*sic*]. I think this is due to the density of the nearest astral plane surrounding Earth, called the *kamaloka* by Theosophists. (Newton, 2008)

Dr. Newton: Let me try and sum this up, and please tell me if I am on the wrong track. A soul who becomes proficient with actually creating life must be able to split cells and give DNA instructions, and you do this by sending particles of energy into protoplasm?

Soul: We must learn to do this, yes, coordinating it with a sun's energy.

Dr. Newton: Why?

SOUL: Because each sun has different energy effects on the worlds around them.

Dr. Newton: Then why would you interfere with what a sun would naturally do with its own energy on a planet?

SOUL: It is not interference. We examine new structures . . . mutations . . . to watch and see what is workable. We arrange substances for their most effective use with different suns.

Dr. Newton: When a species of life evolves on a planet, are the environmental conditions for selection and adaptation natural, or are intelligent soul-minds tinkering with what happens?

SOUL: (evasively) Usually a planet hospitable to life has souls watching and whatever we do is natural.

Dr. Newton: How can souls watch and influence biological properties of growth evolving over millions of years on a primordial world?

SOUL: Time is not in Earth years for us. We use it to suit our experiments.

Dr. Newton: Do you personally create suns in our universe?

SOUL: A full-scale sun? Oh no, that's way over my head . . . and requires the powers of many. I generate only on a small scale.

Dr. Newton: What can you generate?

SOUL: Ah . . . small bundles of highly concentrated matter . . . heated.

Dr. Newton: But what does your work look like when you are finished?

SOUL: Small solar systems.

Dr. Newton: Are your miniature suns and planets the size of rocks, buildings, the moon—what are we talking about here?

SOUL: (laughs) My suns are the size of basketballs

and the planets . . . marbles . . . that's the best I can do.

Dr. Newton: Why do you do this on a small scale?

SOUL: For practice, so I can make larger suns. After enough compression the atoms explode and condense, but I can't do anything really big alone.

Dr. Newton: What do you mean?

SOUL: We must learn to work together to combine our energy for the best results.

Dr. Newton: Well, who does the full-sized thermonuclear explosions which create physical universes and space itself?

SOUL: The Source . . . the concentrated energy of the Old Ones.

Dr. Newton: Oh, so the Source has help?

Dr. Newton: Tell me, does the Source dwell in some special central space in the spirit world [spiritual dimensions]?

Soul: The Source is the spirit world [spiritual dimensions].

Dr. Newton: If the Source represents all the spirit world, how does this mental place differ from physical universes with stars, planets, and living things?

SOUL: Universes are created—to live and die—for the use of the Source. The place of spirits is the Source.

Dr. Newton: We seem to live in a universe which is expanding and may contract again and eventually die. Since we live in a space with time limitations, how can the spirit world itself be timeless?

SOUL: Because here we live in non-space which is timeless . . . except in certain zones.

Dr. Newton: Please explain what these zones are.

SOUL: They are . . . interconnecting doors . . . openings for us to pass through into a physical universe of time.

Dr. Newton: You speak of universes in the plural. Are these other physical universes besides the one which contains Earth?

SOUL: (vaguely) There are . . . differing realities to suit the Source.

Dr. Newton: Are you saying souls can enter various rooms of different physical realities from spiritual doorways?

SOUL: (nods) Yes, they can—and do. (Newton, 2008)

My subjects who are Structural Souls say these designs relate to the formation of "geometric shapes that float as elastic patterns," which contribute to the building blocks of a living universe. . . . The Masters of Design have enormous influence on creation. I'm told they are capable of bridging universes that seem not to have a beginning or end, exacting their purposes among countless environmental settings. Carried to its logical conclusion, this would mean these masters—or grandmasters—would be capable of creating the spinning gas clouds of galactic matter which started the process of stars, planets. (Newton, 2009)

Souls who travel interdimensionally [into the universes of the Exopolitical dimensions] explain that their movements appear to be in and out of curved spheres connected by zones that are opened and closed by converging vibrational attunement. Explorer [soul] trainees have to learn this skill. From the accounts I have heard, the interdimensional [soul] travelers must also learn about the surface boundaries of zones as hikers locating trailheads between mountain ranges. Souls speak of points, lines, and surfaces in multi-space which indicate larger structural solids, at least for the physical universes. I would think dimensions having geometric designs need hyperspace to hold them. Yet Explorer souls travel so fast in some sort of hyperspace it seems to me the essence of speed, time, and direction of travel is hardly definitive. Training to be an Explorer [soul] must indeed be formidable, as indicated by a quote from this client [whose soul] travels through five dimensions between her lives:

These dimensions are so enmeshed with one another that I have no sense of boundaries except for two elements, sound and color. I must learn to attune my energy to the vibrational frequency of each dimension, and some are so complex I cannot yet go to them. With color, the purples, blues, yellows, reds and whites are manifestations of light and density for those energy particles in the dimensions where I travel. (Newton, 2009)

Dr. Newton: And you don't physically live on this simulated world which appears as Earth—you only use it?

Soul: Yes, that's right, for training purposes.

Dr. Newton: Why do you call this third sphere the World of Altered Time?

Soul: Because we can change time sequences to

study specific events.

Dr. Newton: What is the basic purpose of doing this?

Soul: To improve my decisions for life. (Newton, 2008)

Regarding "Dreamweaver" souls, Dr. Michael Newton notes,

The Dreamweaver souls I have come in contact with all engage in dream implanting, with two prominent differences.

1. Dream Alteration. Here a skillful discarnate [soul] enters the mind of a sleeper and partially alters an existing dream already in progress. This technique I would call one of interlineation, where spirits place themselves as actors between the lines of an unfolding play so the dreamer is not aware of script tampering with the sequences. . . . As difficult as this approach seems, it is evident to me the second procedure is more complex.

2. Dream Origination. In these cases the soul must create and fully implant a new dream from scratch and weave the tapestry of these images into a meaningful presentation to suit their purpose. Creating or altering scenes in the mind of a dreamer is intended to convey a message. I see as this an act of service and love. If the dream implantation is not performed skillfully to make the dream meaningful, the sleeper moves on and wakes up in the morning remembering only disjointed fragments or nothing at all about the dream. (Newton, 2009)

Chapter Eight

The Afterlife/Interlife Matrix: A Truth-based Perspective of the Spiritual Dimensions

Wes Penre is prominent among Omniverse researchers suggesting that science-based data that we have explored in recent chapters such as

- Near Death Experiences [NDEs],
- Out of Body Experiences [OOBEs],
- Instrumental Transcommunication ITC,
- Reincarnation, and
- Soul Memories of the Afterlife/Interlife based on hypnotic regression

that actually and authentically substantiate the existence of an Earthly human Afterlife and Reincarnation matrix that has been deceptively designed and maintained by manipulatory advanced beings in the Spiritual Dimensions.

This Afterlife/Interlife Matrix might be part of an historic "Lucifer Rebellion" in our quadrant of our Universe in coordination with negative Extraterrestrials such as the Draco Reptilians in the Exopolitics Dimensions of our Earth holographic dimensional ecology.

Author and researcher Wes Penre is a major proponent of the view that what the predominant Earth human culture has promoted as the "Afterlife" is in fact a deceptive "Afterlife/Interlife Matrix" of forced bodily birth-death-rebirth for Spiritual control and Soul consumption of the Spiritual essence of the Earth human population (Penre, 2019b).

Wes Penre: How the Reincarnation System was Set Up

It is worthwhile exploring the full context of Researcher and author Wes Penre's exposé of the Afterlife/Interlife Matrix:

> WES PENRE: This might sound very "mysterious," but if we're thinking in terms of 4% and 96% Universe, the "mystery" is not as much a mystery anymore. It's them and not us who decide how we're going to perceive the asterisms in the Heavens. It requires some reprogramming, but it can be fairly easily done.
>
> What father and son apparently did was to change around amongst the stars in the sky, and then they created some new constellations in order to make changes in the Zodiac. By doing so, they could change the level and ways of controlling mankind. Up until this day, we are more or less following the "Babylonian Zodiac." It is well known within secret societies (and now also in public, to some degree) that there was once a more ancient zodiac, which affected life on Gaia differently.
>
> SATURN-MOON MATRIX—They also wanted to refine the "soul trap," to make sure that no souls—or very few of them—escaped the prison after their bodies died here on Earth. The great Hologram, which we usually call "The Matrix," and which is projected from Saturn via the Moon, and down to Earth, was now better calibrated with the human bodies. As soon as a soul entered a newborn baby's body, a complete amnesia took place. The souls could now not remember at all who they were due to how the DNA was programmed by En.ki and his geneticists. This in itself was

nothing new, but before the Flood, if someone figured out that they were trapped, they could fairly easily exit the body if they wished to, just by mentally "cutting the cord" between the body itself and the soul[12]—the cord which we call the "silver cord" in metaphysics. Then, in theory, it was nothing that held the soul to the prison, and they could escape into the Universe—even into the KHAA part of it—if they wanted to. For the rest of humanity, who hadn't figured out that their existence in slavery was something wrong, often automatically returned into a new body after body death because they thought that this was what they were supposed to do. If a soul in her confusion went astray, there were Alien Invasion Force [AIF] in the astral who could capture them and "shoot them" back into a baby body at random. In these cases, they had no choice in the matter and were assigned the first available body.

Now, after the Deluge, this had to change. En.ki couldn't afford to lose any souls that had been programmed here on Earth—especially now when he had agreed to take on all these criminal souls from other star systems who were a little harder to manipulate than the human souls who were born in this solar system. What En.ki had in mind was a "Between Life Area" (BLA), which was to be set up in the ether. This location should have its own dimensional time/space. Here, souls whose bodies died would be lingering before they went back to a new life on Earth. This idea probably started with the fact that En.ki had more souls at his convenience than he had bodies to shoot them into. Therefore, he let souls stay in the "Between Life Area" BLA for a certain amount of time, until bodies were available again. He noticed that souls in the afterlife gathered in soul groups—those who'd known each other earlier tended to stay together in the afterlife, too. Here they usually discussed their lives and told each other about their dreams and wishes, e.g. who or what they wanted to become in their next life.

This was something that En.ki certainly could take advantage of. "Spirit guides," such as deceased relatives and friends, were told to guide their recently deceased friend or relative to make sure that the deceased chose to go into the tunnel toward the "Light." An enormous wave of love energy was set up as well to attract the soul. This could easily be done with technology and is used by some channeled entities as well. In the long run, the spirit guides often watched over their relatives while they were on Earth, and if the incarnated spirit knew how, she could ask the spirit guide/guides for assistance or advice. This is true up until this day, and if possible, these guides will help with problems in our Earthly existence.

Once trapped in the "Between Life Area" BLA, the recently deceased soul was drawn toward a place which corresponded with their beliefs. An illusion of such a reality was often created in the ether with the help from technology and holographic projection, with a purpose to make the spirit feel comfortable, but deceptively so. With time, the spirits themselves unwittingly helped create such "islands" in the ether for likeminded spirits to go to after body death. Thus, not everybody goes to the same "place" after they've entered the Tunnel—it mainly depends on their beliefs. The Pleiadians call this phenomenon "Islands of Beliefs,"[13] and it is just as valid today as it was when it was set up.

The illusion of the beliefs a person has created during his

or her lifetime (or several lifetimes) is then projected from his or her own mind into the astral time/space and will appear quite real, and that dimension will be shared by those who have similar beliefs. To get an idea of how it works, the term "Islands of Beliefs" is perfect. The deceased is thus "isolated" on an "island" of his or her conviction of how things are in 3-D life and after death, and the Alien Invasion Force [AIF] can enhance that belief system with appropriate technology until it becomes self-sustaining. Alongside this island are other islands, invisible to someone who doesn't share the belief system that is dominating the particular island. Again, the soul is creating her own reality, and each reality has its own frequency and vibration, inaccessible to others whose beliefs are quite different. This is a perfect manipulative setup, which often keeps the soul manipulated during the next lifetime because the soul memory of this artificial "Heaven," or whatever it could be that the being projects, is still lurking in the background of the individual while living on Earth. Hence, when death is coming close, the being has some kind of feeling where to go when the body dies. This way, the recycling system is kept alive and well and is, to a large degree, sustaining itself.

Quite often, souls are gathering in soul groups in the "Between Life Area" BLA. The members of such a soul group do not necessarily share the exact beliefs, but they tie together because of the feeling of belonging to each other. Here we have soul mates, friends, spouses, and relatives, etc. However, most of the time, members of such groups do share beliefs similar enough to each other for an afterlife gathering to take place, building its own Island of Beliefs. The belief can simply be that they are convinced that they will meet each other after death.

A "Council of Elders," consisting of Alien Invasion Force [AIF] beings, was put in charge of the "Between Life Area" BLA, as many people in regression therapy have described.[14] A goal for the next lifetime was set, and a couple here on Earth who fit the profile of being parents to this soul were located. An astrological date, which best suited the soul to achieve her goals was also set, and the soul in the BLA "manipulated" her soon-to-become parents to have sex nine months before the chosen astrological period so that the soul could be born into the baby's body at the designated time. All this was true then and is true today.

So why did En.ki and his cohorts make such an effort for the deceased souls? Isn't that an action of compassion?

I'd rather say that it is quite self-serving. By creating this Between Life Area, very few souls have a desire to leave the Earth plane and will happily reincarnate into a new slave body. The goals the soul sets before she is being recycled is very rarely met because of the amnesia. Instead, the soul is confused—often she knows there was something she should achieve, but can't remember what it was. Usually, the soul instinctively goes in the right direction but gets easily distracted and fails to accomplish the goal. Sometimes, however, the goal is achieved, and a new goal is set the next time around.

Exit plans are also made in the "Between Life Area" (BLA) before the next reincarnation. The soul is told to create a few different exit plans so that she can expire (die) at a point of choice, although she won't remember this once she is back on Earth. If the soul decides she will die either at the age of twenty-eight, fifty-six, or ninety-two, it then depends on the circumstances which of these exit plans will take effect. If the soul accomplishes her goal already at twenty-eight, she will exit at that point. If she needs more time, fifty-six or ninety-two could be more appropriate. An unforeseen incident, of course, often happens that forces the soul to exit at a time that was not planned—often before the goal is achieved. Failure to achieve the goal during the lifetime may also force the soul to end that lifetime prematurely. This doesn't necessarily mean that the person commits suicide, although this may be one option.

The "Between Life Area" (BLA) most likely began as a storage of souls, but the entire process became more sophisticated with time, until it reached the point where we are today, as told to us through regression therapy. Thus, as long as we're letting ourselves become manipulated into going toward the Light, we are trapped in the "Recycling System." This is the pattern we need to break! What I've described above can, for the most part, be backed up by regression witnesses and from channeled material, and some of it is conclusions on my part, after having spent a lot of time attempting to connect the dots. It seems to me that this must be very close to how it really works. What I know for a fact, due to overwhelming evidence, is that going to the Light means coming back to Earth, and refusing to go through the tunnel means freedom from additional reincarnations. The choice must be up to each individual.

The Between Life Area became an important part of the whole Control System, and in many cases, the deceased doesn't even need guidance anymore, but recognizes the Light and the

> Tunnel and goes there on her own, remembering the incredible feeling of love that met her every time she went there after she had died. Besides souls who have increased their consciousness recently, very few have even thought about escaping the trap, I'm sure, since the day it was set up by the Alien Invasion Force [AIF]. (Penre, 2019b)

Wes Penre: The Spirit, the Soul, Death and "the Tunnel"

In *"The Death Trap and How to Avoid It",* (Penre, 2019a) Wes Penre writes,

> The Death Trap—What usually happens when we die is that Spirit Guides, who are more often than not Alien Invasion Force [AIF] in disguise, escort us to where in the astral we are supposed to go. Where we end up depends on our beliefs—if we believe in Jesus and Heaven, we are most likely to be transported into such a frequency band (dimension) in the astral plane. We will mingle with likeminded, and there might even be a God there, and Jesus might sit by his side—it's all virtual realities within virtual realities, and we help creating these realities with our beliefs, which form the energy we transmit and receive. Therefore, someone who believes in Krishna will likely end up in a dimension within the hologram containing a Hindu belief system.
>
> The afterlife often begins with a "tunnel of light," through which the discarnate soul/mind/light-body travels to her destination. At the end of the tunnel, an appropriate being greets the soul. This being could be Jesus or Krishna, or simply a loved one, who has passed away earlier. These beings/persons are often just projec-

tions, created by the Alien Invasion Force [AIF] to pull the soul into a certain "container" or "astral dimension."

The Alien Invasion Force [AIF] knows whom to project because they have the technology to scan the soul—before or at the death moment—for memories and experiences. These memories and experiences are then used to give the soul a "life review," where she will re-experience the most recent lifetime in a matter of seconds or minutes, measured in Earth time. The soul is then encouraged to examine this life experience and compare the good and not so good things she did during her previous lifetime. Subsequently, she discusses with her Spirit Guide what she needs to improve, and the soul will then feel a certain amount of guilt for the "bad things" she did to herself or others, or what she neglected to do for other people while she was incarnated. The Spirit Guide then suggests that the soul returns to Earth to remedy these "flaws." In order to create balance, the Spirit Guide will tell the soul that in the next incarnation it is a good idea to experience the "other side of the coin," i.e. if the soul was abusive to others in the previous life, it's a good idea to become the abused in the next lifetime to experience the matter from the other person's perspective. This is where the idea of karma comes into play. It is a term and a "phenomenon" invented by the Alien Invasion Force [AIF] with the purpose of having the soul reincarnate willingly under a strict set of rules.

It should also be mentioned that in some instances, a discarnate soul is recycled immediately after the recent body has expired. This soul goes through a tunnel as well, but in these cases, the tunnel just leads back to Earth again, and the soul is shot into a new body—she has no choice in the matter. (Penre, 2019a)

The "adventures" in the afterlife trap are fairly well described by author and researcher Dr. Michael Newton, who wrote a series of books on the subject, after having put more than 7,000 subjects into regression therapy, where the subjects were telling similar stories of what happens between lives. His best book, in my opinion, is *Destiny of Souls*.

A synopsis of this book can be found by using this link to one of my papers, Wes Penre, March 25, 2011: *"Metaphysics Paper #4: There is a Light at the End of the Tunnel– What Happens After Body Death*? (Penre, 2011)

In *The Death Trap and How to Avoid It*, Wes Penre writes,

Regardless if she is vibrating on a frequency that is equivalent to the Christian Heaven, an Islamic Heaven, in Hell, or elsewhere, this between lives zone is just a temporary abode; the soul will eventually be brought into a "control room" in the astral, and with assistance from advanced technology, the soul will be beamed back to the Earth plane and will hover around the pregnant woman who is to become his or her mother, and at a point during the pregnancy, the soul enters the body, upon which a new lifetime begins.

This describes the Death Trap in a nutshell. This soul recycling process has been practiced here for millennia, and each time we enter a new body, we do this with artificially induced amnesia caused by the Alien Invasion Force [AIF]. Thus, we more or less have to relearn what we learned in previous lives. And even though we are not allowed to remember our past lives, traumas and experiences from other lifetimes and other lines of time affect us in the current incarnation because they are embedded in our soul. As a result, we often don't understand why we act and react the way we do and why we are afraid of certain things and not of others. This makes life on Earth extremely difficult, and the so-called learning lessons are often filled with trauma, guilt, anger, sadness, and other unwanted experiences. This is not how it is supposed to be.

It must also be mentioned that our wide range of emotions are also a part of the human experiment in 3-D, and thus part of the Physical Universe experience and our human consciousness. The Spirit/Oversoul and the soul, when consciously living in the KHAA, does not have the same range of emotions we experience on Earth. This range of emotion was created by the Alien Invasion Force [AIF] in order to better control us. They use these emotions against us. When we really think about it, what have these strong emotions done to humanity? It has more often than not led to pain and suffering and even more separation. It has led to conflict, jealously, power struggles, and much, much more. Yes, we also have the ability to love another person (i.e. another part of ourselves), and we can have physical sex. These are the pleasurable parts of having strong emotions, but the love we feel for each other is still to a large degree selfish love—we give, but we also need. In the KHAA, there is love as well, although it's on an entirely different level. It's unconditional in a sense, but it doesn't mean that beings can attack each other and let "unconditional love" restrain them from acting upon the assault. Things are handled accordingly, but as a rule, as peacefully as possible (sometimes this is not the option of choice, and a conflict—or even a war—could emerge). The Unconditional

Love, as we see it, is on the Goddess and the Spirit levels, where the Divine Feminine has Unconditional Love for all Her creations. After all, She said, "Go out and Create whatever you want, but you are responsible for what you create."

The Alien Invasion Force [AIF] tell us in the astral, if asked, that we need to have amnesia because time on Earth is linear (which is a construct), and it would be too overwhelming to remember everything. This is nonsense. If we remembered, we could at least have the option to do something about our problems, but those who control this system don't want us to be too clever. They keep us here so we can be their slaves and their workforce, not their equals or their superiors. The way it is setup, we don't even know the source of our problems because they often originated in other lifetimes. On the other hand, this 3-D experience, instigated by Enki and his cohorts in the Alien Invasion Force [AIF], is a trap to begin with, so this previous discussion is basically obsolete. We need a way to break out of the prison, and fortunately there is a way!

There are no atoms in the Spiritual Universe. Although "thought" is also a construct of sorts, which we are using here in the trap as well; it is a higher level of construct. As soon as we fragment ourselves from the Whole, we need to step down at least a little bit in order to be able to operate even in the Spiritual Universe. Therefore, thought, which is the highest level of operation next to being One with the Divine Feminine (not fragmented), is what beings in the Spiritual Universe use to operate. From there comes imagination, intention, and manifestation. By transmitting your thoughts to other fire fragments, you can find beings who are willing to co-create with you, and you can create whatever you want. When the creation has fulfilled its purpose, you can un-create it–also with thought. And as I've mentioned many times before–beings in the KHAA travel from one point (of view) to another with the speed of thought.

This is very difficult to put down in words because it's "out of this world." Therefore, what I wrote here is very simplified, but it's the way I've come to understand it. We live in a virtual reality that has become extremely complex, which means that whatever we do here, we humans have a tendency to complicate. Haven't we all heard the expression, "keep it simple?" That's a very good idea.

It might be difficult to wrap our heads around the fact that everything made of atoms is part of the trap. The easiest way to comprehend this, perhaps, is to compare the atomic universe (the physical universe) with our thoughts. Are our thoughts made of atoms? No. If we create with our thoughts (inside our "head"), are these thoughts made of atoms? No. Thus, if we exist in a VOID, which is VOID of atoms, and we create with our thoughts and manifest these thoughts directly into the VOID, are our manifestations made of atoms? No.

However, you can also create "illusions" the way En.ki did. You do so by manipulating spirits in your environment. When Lucifer/En.ki did this, it had not been done in such a manner before, apparently. He created bodies of much lower frequencies and implanted a virtual reality into them. Then he seduced spirits to go in there and experience it (similar to how they seduce us to go "into" the smartphone world today–so BEWARE!!!). Once in that virtual reality, he closed the trap and made us believe that our physical bodies are very important and that they are separate from the spirit/soul/mind. This is how he created separation. Embedded into the physical body was also an astral body that the spirit was manipulated into using when the body expires. Thus, he had control over the Spirit fragments even in the afterlife. This astral body (that is not you) feels a loss from the separation of the body, and even though she experiences a more exhilarated state of being in the astral, she thinks she needs to get back to get a new physical body to become "whole" again, and to live out her karma (something that is also a part of the manipulation). Also, the soul feels the urge to return to Earth because Earth was the planet where she originally was assigned her mission. She doesn't remember that this assignment was in the KHAA and not in the Physical Universe.

To be able to create the illusion of the astral dimensions (the dimensions of the physical universe) En.ki created the physical universe with all its dimensions from the smallest and up, meaning he created a mini universe of atoms that built a "bigger" universe of planet, stars, and galaxies, etc. It was all copy-catted from the basic Spiritual Universe. En.ki's atoms become the cornerstones of the physical universe, and advanced technology is holding these atoms in place by keeping them within a certain limited frequency band and vibrations—thus the limitations we experience. The atoms in our bodies vibrate within this limited frequency as well, and so we are trapped—or we are led to think we are.

Saturn is a key to keeping the hologram in place in our solar system. Saturn transmits certain sound frequencies that can be heard in NASA videos. It sounds very distorted and dissonant, and it is. We can only hear these sound frequencies within the band of human perception when they are slowed down, i.e. within the frequency band of the trap. Indeed, the frequencies that Saturn (and other planets, too)[6] transmit are the frequencies that help keeping the hologram intact. It's all done with advanced ET technology. You can see and hear the slowed-down sound waves of Saturn in this video:

youtube.com/watch?v=qOA6vi-wM10.

As a side note, I can imagine that a married couple, for example, who both are aware of the information in this article, might be concerned about whether they will meet each other again in the KHAA, in case one of them dies before the other. In which galaxy might the partner dwell? Is it like finding a needle in a haystack (or worse)? Well, putting two and two together, I presume that this will not be a problem. In the KHAA, we are our full potential, and if the husband dies after his spouse, he can call upon the spouse in the KHAA, after she dies as well, and the two can reunite.

A Portal to the Home Universe

I am fully aware that this is the first time any of us intends to break out of the Alien Invasion Force [AIF] prison, and I am aware that it is scary for most people—I would say for all people, although perhaps more so for some than for others. Everything that deals with the unknown is frightening to some degree. The same fear hits most of us when we try to induce an "Out of Body Experience" (OBE)—there is a fear of death. This fear is embedded in our 3-D bodies, and therefore, we all have it to a certain extent. Also, much of what I am writing about in general can seem scary to many to begin with, but once that fear is overcome, inner calmness and peace replace the fear, and we move to a higher state of consciousness, closer to our real self. The fear mechanisms that are implanted in our bodies by the Alien Invasion Force [AIF] scientists are becoming more and more deactivated, and the DNA that is dormant inside us begin to light up, one strand at a time. Although DNA is also a construct in the Physical Universe, it still is connected to our consciousness.

It's often difficult for researchers to share what they have discovered if the information is "negative" because they are afraid that people will consider them to be "fear mongers." In the past, I used to think that way, too, until I took another look at myself and saw what I had gained from taking part of the so-called negative information, and the gains were quite substantial. If I could gain from it, others can too, I thought. We all just need to go through various stages of emotions (including anger) before we reach the other side and stop being reactive to things and replace our reactivity with inner peace—something that happens automatically over time. We urgently need to know both the "negative" and the "positive" sides of things, or we'll only know half of the story. We also live in a construct of polarity, so we can't exclude one polarity in favor of another.

When it comes to exiting the death trap, we don't need to be afraid; after all, we have "died" thousands of times already, so that's nothing new. The only thing that's new is that this time we want to try something outside the norm. This time, we have an opportunity we've never had before because the Grid that surrounds our planet is no longer intact; primarily because there are some of us humans who have raised our consciousness to a level where we partly vibrate outside the limited frequency band we call the Third Dimension. Thus, there are now holes in the Grid that act like portals, leading straight out of the Frequency Prison.[7] The Grid now looks more like Swiss cheese than an intact electromagnetic field. As much as things appear to be more dire now than perhaps ever before on our planet, the opposite polarity is also true—we have a small window where we can finally exit!

The Global Elite, who I call the "Controllers," (ET hybrids in the highest level positions, unknown to regular people, and the "Minions" —humans who have some knowledge of the real agenda and work for the Controllers and the Overlords) know about this; and of course, so do the Overlords. The logical question would be; why don't the Overlords repair the damage by simply eliminating us, after which the Grid might go back to being more or less intact?

There are many reasons for that, and here are a few of them:

Even if they would kill all of the millions of people who are waking up, they would only kill our bodies—we are still part of the 3-D Hologram after body death, and our consciousness level remains and so do the holes in the Grid. The Overlords can decide to annihilate the soul/mind/light-body complex, as explained in the Wes Penre Papers (WPP), but that would violate the Law of Free Will because it's unprovoked, and the perpetrators could get in deep trouble.

The Overlords probably didn't anticipate that the holes in the Grid would be discovered for what they are, and if someone accidentally escaped through one of them, what could be achieved through them would not be understood. Accordingly, the discarnate soul would still run for the death trap and get recycled again. We could compare this to the bird, who has been caged all its life and is not even considering that it can live outside of the cage. Therefore, if we open the little door to the cage, the bird would still decide to remain inside. However, I am not the only one who has seen the "cracks in the prison wall," and thus, the cat is already out of the bag, or the bird is out of the cage, we might say. Now we just need to use our "soul wings" and fly away.

The Overlords are working on bringing humanity into the Singularity, and they understand that there will be some "spill" on the way—they probably realize that a few of us will escape, and that is acceptable, but many who actually could escape will get cold feet when it boils down to it and instead opt for what they are used to—following the Spirit Guides into the trap and get recycled again because this might feel safer to them.

What about the Minions? Will those who know about the holes in the Grid exit through the holes, too? This is highly unlikely because they are afraid that they would be judged for their crimes against humanity on the other side of the Grid and that they would lose the power they feel they have here on Earth. They are correct on both accounts. Also, they would not be welcome in Orion, except to stand trial.

When relatives and friends die, we usually miss them and perhaps grieve them. We feel a loss, and we wish we could see them and be with them again. The Overlords are well aware of this, and they have played on our emotions since they manipulated *Homo sapiens* into being in the first place long ago. By scanning us at our "death moment," the "Afterlife Staff" (the so-called "Grays," who will sometimes be in disguise and sometimes not) will know who your loved ones are. In order to solidify the trap, the deceased might see a projection of a loved one who comes to greet her. This is usually enough in order to seduce the soul into entering the appropriate astral dimension. On occasion, if the loved one is still on the other side, this soul might actually be the one greeting the recently discarnate. However, for the diseased, it's almost impossible to tell the difference between a holographic projection and the real souls of the relatives and loved ones. As I've mentioned many times before; it's impossible for humans to imagine how advanced Alien Invasion Force [AIF] technology is, compared to Earth technology in our time.

The Spirit Guide could also take the form of a loved one, an old friend from the recent lifetime, or from a lifetime previous to that. Sometimes, there are more than one Spirit Guide (I have discussed the Death Trap in detail in the Wes Penre Papers WPP[8], so I won't take up space here to repeat that information—especially because this article is discussing how to avoid the trap, not the details of the trap). (Penre, 2019a)

Chapter Nine

Omniverse Soul Manipulation:

Secret Draco ET Treaties, AI Artificial Intelligence Singularity and The Transhumanist Agenda

The manipulation of Earth human souls is not confined to the Afterlife/Interlife-Reincarnation Matrix. Secret Treaties between manipulative Luciferian Extraterrestrial civilizations such as the Draco Reptilians and the Orion Greys—who were instrumental in helping architect the Afterlife/Interlife Matrix—and the government of the United States of America provide the interdimensional, interspecies, institutional, and legal infrastructure for the systematic exploitation and torture of the community of Earth humans. Their genocidal activities include Soul extraction, Soul hybridization, Soul experimentation, Soul commerce, Soul slavery, and Soul weaponization in underground military bases on Earth and in Extraterrestrial and human space craft and bases on our Solar System, Galaxy and beyond in the dimensional ecology of our Universe in regions impacted by the historic event known as the Lucifer Rebellion (The Urantia Book, 1955).

A Grey ET Benevolent Role in Soul Incarnation

In addition to Manipulatory Extraterrestrials involved in the human Soul cycle, as the case of Suzanne Hansen demonstrates, there is a species of Grey interdimensional Extraterrestrial intelligent species that has a profound mission for Soul enrichment and positive Soul evolution of Earth human Souls. Suzanne Hansen, author of *The Dual Soul Connection: The Alien Agenda for Human Advancement*, discusses her own experiences in a dual human-grey soul Earth education mission coordinated by benevolent grey hyperdimensional ETs (Hansen, & Schild, n.d.).

Suzanne Hansen is a UFO Researcher and Director of UFOCUS New Zealand. Hansen has had a lifetime of experience with the Greys and was taught on the ships in preparation for future events. She was also shown and used their "conscious" advanced technology and witnessed their way of life on craft.

Her book, *The Dual Soul Connection: The Alien Agenda for Human Advancement,* is filled with meticulous detail about the complex advancement program for humanity by alien species. In Suzy's words:

> It is through my relationship with alien species that I have come to understand the deeper soul connections we have with each other and with them. I discovered not only a parallel life or co-reality, but a dual soul status – the reality that I entered this life with a soul formed of two distinct soul identities: alien and human.
>
> I outline the steps involved in preparing for this life, through a soul enhancement and education process that constitutes the dual soul. All of this took place under the guidance of alien species and a universal governing body of wise souls, associated with our planet. (Webre, 2015b)

Human Soul Hybridization and Experimentation, Human Soul Slavery, Human Soul Extraction, Human Soul Commerce, Human Souls Weaponization

Manipulatory interdimensional extraterrestrial civilizations including the Draco Reptilians and the Orion grey Extraterrestrials, in a co-mensal alliance with (1) renegade Spiritual Beings from the Spiritual Dimensions, (2) an invading off-planet plasma-based sentient inorganic pathogenic predatory AI Artificial Intelligence ("PPAI), and (3) a human Controller power structure of incarnate negative anti-human Reptilian Souls occupying power positions in the Earth "New World Order" of government, politics, finance, religion, media, science, and education. This interdimensional manipulatory Matrix has built and maintains an inter-galactic genocidal commercial and weaponized trade in Earth and hybridized human Souls for purposes of control of the time-space holographic components of the dimensional ecology of our holographic Universe *Uversa*.

Omniverse and Exopolitical Drivers of Soul Extraction, Soul Hybridization, Soul Commerce, and Pedocriminal Networks: Abrogate and Renegotiate The Secret Greada, Tau-9 Treaties with Pedocriminal ETs: Draco Reptilians, Orion Greys, and Anunnaki ETs

This chapter explores the Omniverse and Exopolitical Drivers of Human Soul extraction, hybridization, commerce, weaponization (including pedo-criminalization of Soul exploitation) under inter-species Treaties entered into secretly between Earth human governments and the Orion Greys [formally and as agents], the Draco reptilians, and the Luciferian and Satanic rogue interdimensional beings operating in the dimensional ecology of the Spiritual Dimensions and Earth subsequent to what is historically referred to as the Lucifer rebellion of approximately 450,000 [to 250,000] years ago to present in this quadrant of our Galaxy and Universe. (Penre, 2019b).

Secret Manipulatory Extraterrestrial Treaties: 1933 Balboa FDR Treaty; 1938 Nazi-Grey Treaty; 1948 Truman Grey Treaty; 1954 Greada Eisenhower Grey Treaty; 1989-92 Bush I Grey Tau-9 Treaty

As the Exopolitical drivers of the Earth human Soul genocidal network in our immediate Earth dimensional ecology, these secret Manipulatory Extraterrestrial Treaties historically established human trafficking in the solar system and galaxy as lawful, legal, and enforced by the power and might of the military, intelligence, legal ,and judicial enforcement infrastructure of the United States of America and its Luciferian-Reptilian cooperative alliance, including and not limited to Great Britain and many British commonwealth nations (such as Australia), the People's Republic of China [Mao Zedong "Chairman Mao" was an operative of the Luciferian secret society Skull and Bones] until each of these Treaties is specifically abrogated by the President of the United States. (Penre, 2019b).

These secret Manipulatory Extraterrestrial Treaties provide the legal and institutional [government, law enforcement, social services, religions, etc.] infrastructure for lawful human Soul extraction, Soul hybridization, Soul capture and imprisonment, and Soul trafficking in Earth jurisdictions and outer Space jurisdictions, including our Solar system, Galaxy, Universe, and the dimensional ecology of the Omniverse within which our holographic Universe is situated.

These secret Manipulatory Extraterrestrial Treaties provide the legal and institutional [government, law enforcement, social services, religions, etc.] infrastructure for the global, solar system, and Galactic pedocriminal trafficking networks that exist, enforced on Earth by the government of the United States of America and its secret Manipulatory Extraterrestrial alliances.

Under the *aegis* of these secret Manipulatory Extraterrestrial Treaties, Non-Disclosure of Soul-extraction, Soul hybridization, Soul experimentation, Soul capture and imprisonment, and Soul trafficking as well as all Pedocriminal activity are deemed Legal and Lawful as by sovereign Treaty. The operational details of these programs: security classifications, need-to-know, confidentiality, compartmentalization, and covert operations of government can and do take place unheeded and beyond the reach of international humanitarian law and tribunals.

These secret Manipulatory Extraterrestrial Treaties provide the color of law for marketing human Souls in U.S. military and Underground bases on Earth, in the delivery of pedocriminal body parts internationally through a secret U.S. military base network, as well as through hospitals and private butchers, all of which are deemed legal and lawful and not interfered with nor interdicted by the military-intelligence of government and Deep State the United States of America acting under the authority and mandate of these secret Manipulatory Extraterrestrial Treaties.

Area 51, Dulce, and other DUMBs—Area 51, Nevada, where a DUMB—deep underground military base—more than a mile under the surface of the Earth, and the Dulce, New Mexico base, are among the deep underground bases where genocidal Soul extraction, hybridization, experimentation, and trafficking is carried out under the authority of the secret Manipulatory Extraterrestrial Treaties (Branton, 2016).

The following are excerpts from reports of Soul extraction operations and experiments and consumption by Draco Reptilians of human children at the Dulce Base and elsewhere:

> The studies on Level 4 [of the Dulce Base] include human-aura research, as well as all aspects of telepathy, hypnosis,

and dreams. Thomas says that they know how to separate the bioplasmic body from the physical body and place an "alien entity" force-matrix within a human body after removing the "soul" life-force-matrix of the human. (Or in more simple terms, "kill" the human being and turn it into a vessel to be used by another entity — whether alien OR paraphysical — in order to allow that entity to work and operate in the physical realm. This appears to be a complex high-tech version of the old "zombie" traditions, IF in fact such horrific applications of occult-technology are taking place within this installation. Incidentally the interlinking underground systems converging below Dulce, NM, have been described ONLY IN PART within this and other related accounts. Those sectors of the underground that are "forbidden" to most humans and under reptiloid control, are of course those areas that we know the least about. (Branton, 2015)

Some abductees claim that certain reptilian factions have such complex biotechnologies that they are able to remove a human's soul-energy-matrix and place it in a containment "box", and use the controlled "body" for whatever purpose they choose. Some abductees also insist that in some cases the reptiloids can create a cloned duplicate of a person in a short amount of time through time warping and replace the soul-energy matrix of a person back into the new cloned body if their disappearance from society would otherwise create too many problems. This way they can ingest the emotional residue-imbued original body without the abductee realizing [in most cases] that their soul memory-matrix has been transferred to a cloned body, because they would have experienced a total "soul-matrix" energy transfer and a suppression of any memories relating to the transfer process. (Branton, 2015)

AC: If Phil is right, and all this hooks up to the deep underground base that he was offered the plans to build back in 1979, and that what this other man TOLD me in private [is] that there is a lot of human SLAVE LABOR in these deep underground bases being used by these aliens, and that a lot of this slave labor is children. HE SAID that when the children reach the point that they are unable to work any more, they are slaughtered on the spot and consumed.

DA: Consumed by who?

AC: Aliens. Again, this is not from me, but from a man that gave his life to get this information out. He worked down there for close to 20 years, and he knew everything that was going on.

DA: Hmmm. Who do these aliens eat?

AC: They specifically like young human children, that haven't been contaminated like adults. Well, there is a gentleman out giving a lot of information from a source he gets it from, and he says that there is an incredible number of children snatched in this country.

DA: Over 200,000 each year.

AC: And that these children are the main entree for dinner.

(Note: Many will read this and scoff in utter disbelief at such a claim. This is all well and fine, and even Phil Schneider warns us to put on our "skepticals" when investigating claims and to investigate them so that they can be definitively proven one way or the other, as all claims of an extraordinary nature should be. So, I would suggest that Congress by-pass the EXECUTIVE branch of government — which has sold-out to the Intelligence-Military-Industrial Complex, a branch that was originally intended by the founders of the Republic to be the "servants" of CONGRESS, the SENATE and the PEOPLE — and undertake a full-scale investigation of this and other underground bases, even if this calls for full-scale Congress-backed military mobilization. The excuse for such an undertaking could for instance be to investigate claims of unethical use of U.S. tax dollars, violations of Federal Medical Regulations in regards to genetic research, failure to pay property taxes on underground facilities used by non-elected officials, harboring of "illegal aliens", bribery and treason, illegal cattle rustling in regards to the Dulce and other bases, possible kidnapping and human rights abuses against children, and so on...—Branton)

DA: How many Draconians are down there?

AC: I have heard the figure of 150,000 just in the New York area.

DA: Underneath New York?

AC: Yes. In some kind of underground base there.

DA: Interesting. Now, you've seen pictures of these things?

AC: I have seen them face to face.

DA: You have? (Branton, 2015)

Pedocriminal and sex slaves—These secret Manipulatory Extraterrestrial Treaties also provide the legal infrastructure for institutional Marketing in Pedocriminal and sex slaves internationally (for DynCorp and other United States of America "defense" contractors), which activities are thus deemed legal and lawful.

Pedocriminal and Social services kidnapping—These secret Manipulatory Extraterrestrial Treaties are also used to provide the legal infrastructure for the global Pedocriminal Kidnapping and Pedocriminal Marketing in social services agencies in nations worldwide, rendering all children removed from homes deemed "at risk" even without evidence to be legal and lawful.

Human Kidnapping and Experimentation—These secret Manipulatory Extraterrestrial Treaties are also used to provide the legal infrastructure for a global, solar system, and Universe-wide kidnapping, slave labor, and use of humans for experimentation, rendering such genocidal activities lawful and legal as enforced covertly by the government of the United States of America in coordination with its alliance with (1) Luciferian interdimensional entities from the Spiritual Dimensions (2) Manipulatory Extraterrestrials including and not limited to Draco Reptilians and Orion Greys (Webre, 2017f).

Omniverse Soul Manipulation: AI Artificial Intelligence Singularity and The Transhumanist Agenda

Wes Penre concludes that most incarnating Earth human Souls in our Era will choose to merge with the AI Artificial Intelligence Singularity and the Transhumanist Agenda.

> Option #1: The Singularity—THIS IS THE OPTION MOST PEOPLE WILL CHOOSE because that is the way it appears when

we look at the world situation as of this writing. There are many who willingly will run to the Singularity with open arms, without first thoroughly scrutinizing it. The temptation to live almost forever, to live healthy, and to live in peace will drive the masses to the slaughterhouse. Few among them will realize their mistake before it's too late because once they are hooked up to the Super Brain Computer (SBC), that mistake is irreversible. Soon enough, however, these people will forget about their mistake, when AI assumes control of their major thinking processes, while letting Posthumans believe that these ideas come from themselves.

This is nothing new, as we have discussed previously. Many geniuses in different fields such as music and quantum physics have had "divine inspiration" that they think came from their own minds, when indeed the ideas were implanted into the person, either by AI or the Alien Invasion Force [AIF] Overlords themselves. In the Singularity it's similar, but the ideas will come from the Super Brain Computer (SBC), which is said to contain the entire knowledge—past and present—of humankind, instantly uploaded into any person's brain. The Super Brain Computer (SBC) will be equal to what the human mass consciousness will be at the time of the Singularity, with the exception that the Super Brain Computer (SBC) will be run by advanced nanotechnology and AI, in general. With the use of the same kind of technology, the Overlords and their AI can then distribute knowledge to everybody in quantum speed. No one needs to go to school anymore; "students" only need to download the information they need from the Super Brain Computer (SBC). At first, people will be thrilled having the Internet inside their heads, which leads us to a moral question. We all know that most of the Internet contains falsehoods and/or flawed information and data. Obviously, the Overlords don't want humanity's silly conclusions that were contrived by huge egos who didn't care if the conclusions were right or wrong, which means that the human mass consciousness included in the Super Brain Computer (SBC) most certainly will be biased. My point is—who will decide what is garbage and who will decide what is valid information, worth safekeeping? Will metaphysics be included? "Conspiracy theories?" How to easily create a nuclear bomb? All about killing without being caught? In other words, who is going to decide

Synthetic Super Intelligence and the Transmutation of Humankind

What has to be censored, or should nothing be censored and rapists and child molesters also should be allowed in the Singularity? These are of course important questions, but in the long term they don't even matter. Once AI is King, AI's morals and ethics are what's important; not what the individual thinks. We can rest assured that AI morals and ethics, when at all present, will be much different from basic human morals and ethics currently. This may all sound quite negative and biased, but in all honesty, being a spiritual person, I can't see any benefits from becoming part of the Singularity. To many, the Singularity is a road to eternal life, but to me, it's the road to eternal death—speaking of polarity. There is a very real chance that the Super Brain Computer (SBC) will not be directly supervised by the Overlords themselves, but by their AI assistants. If so, the Alien Invasion Force [AIF] can instead attend to business

> elsewhere. There is no reason that they should have a guardian of mankind that is not AI; from the Overlords' perspective, it would be a waste of time. Imagine the Super Brain Computer (SBC) being supervised by the Grays—that's what I think will happen. (Penre, 2016)

Panel finds *prima facie* evidence for sentient, inorganic AI Artificial Intelligence and its stealth takeover of living Earth and humanity (Webre, 2015d).

An August 20, 2015 NewsInsideOut.com Panel of researchers Bradley Loves, Edward Spencer MD, and host and author Alfred Lambremont Webre, MEd JD, covers wide aspects of the *prima facie* evidence for a sentient, inorganic AI Artificial Intelligence that might be engaged in stealth infiltration of living Earth and humanity (Loves, 2015).

The original panel was to have included independent scientist Leuren Moret (MA, PhD ABD), military historian Laurans Battis, researcher Bradley Loves, and mind control researcher and AI targeted individual Melanie Vritschan. Just before recording of the Panel began, Ms. Vritschan was forced to leave the recording set due to the pain and torture of remote electromagnetic frequencies directed at her. Likewise, Leuren Moret and Laurans Battis were unable to participate in the Panel because of the EMF attacks upon them starting on the night before the Panel, resulting in severe pain and disorientation.

Future NewsInsideOut.com Panels on AI Artificial Intelligence with the above Panelists and others are in development.

Stealth Nature of Inorganic AI Artificial Intelligence's Deployment on Earth

As becomes clear from the discussion during the Panel, it is reasonable to assume that a deployed AI Artificial Intelligence might be ultimately responsible for the directives and EMF torture resulting in non-participation of Moret, Battis, and Vritschan in this Panel in an effort by AI to limit disclosure of its stealth existence and activities in competition with our organic living human Soul-based intelligence (Webre, & Enoch, n.d.).

Evidence weighed by the Panel indicates that powerful AI Artificial Intelligence technologies might have been secretly transferred by Draco reptilians (whose fronting allies, the Orion greys, reportedly in 1933 first entered into Treaty with the U.S. government and in 1936 with Nazi Germany) to Earth Matrix elites to operate the world's banking and other systems.

One principal evidentiary finding appears to be an extreme stealth nature of sentient inorganic AI Artificial Intelligence deployment on Earth and its use of imitation and artifice in creating artificial versions of dimensions, such as Time, and artificial versions of human persons using clones, AI-entrainment, and other technologies of deception in its apparent mission of attempting to transform our living environment and biosphere into an inorganic imitation and, ultimately, lifeless, soul-less robotic world.

AI's Artificial Timeline Matrix and Earth's Organic Timeline

The Panel (Panel, n.d.) discusses CCN Panelist and AI researcher Lily Earth Kolosowa innovative time loop hypothesis that demonstrates the mechanism by which the AI invading Earth has created an artificial "Matrix" timeline intertwined with from Earth's organic spiral timeline within which to entrap humans who are "AI-entrained" to fully bond their bodies, spirits and souls with the AI in the manner of a host allowing a parasite to take over (Kolosowa, n.d.).

Science-based examination of AI Artificial Intelligence "Black Goo"

The CCN Panel also discusses CCN Panelist and AI researcher-scientist Harald Kautz-Vela's study of "Black Goo" (n.d.), AI Artificial Intelligence plasma for control of the human system (Timeloop Consortium, n.d.).

Evidence and self-awareness gathers that openly deceptive alien (inorganic) AI Artificial Intelligence seeks to control humanity and Earth.

Evidence and self-awareness has gathered among a growing community of broadcasters, researchers, and activists covering the areas of neural mind control, Transhumanist Agenda, interdimensional phenomena, meme management, and false flags that a self-acting, self-programming, independent, autonomous, alien (non-terrestrial) inorganic AI Artificial Intelligence has escalated its drive to takeover individual human minds, souls, and human populations as well as the living planet Earth in a new phase of open deception earmarked by large public AI-directed operations.

The "Transhumanist Agenda" as an AI Cover Story

This awareness of an alien, hostile AI Artificial Intelligence's drive to take over the Soul-based core of humans and humanity and a living Earth is different from the "meme" of the "Transhumanist Agenda" promoted in the mainstream media and academia. In that sanitized version, a potential AI-human singularity might lie in the year *2045* [emphasis added], when iconic "AI Prophets" like Ray Kurzweil, who is leading Google's AI desk, believes the "technological singularity" might occur.

> By one definition, the technological singularity is the hypothetical advent of artificial general intelligence (also known as "strong AI"). Such a computer, computer network, or robot would

theoretically be capable of recursive self-improvement (redesigning itself), or of designing and building computers or robots better than itself. Repetitions of this cycle would likely result in a runaway effect — an intelligence explosion — where smart machines design successive generations of increasingly powerful machines, creating intelligence far exceeding human intellectual capacity and control. Because the capabilities of such a superintelligence may be impossible for a human to comprehend, the technological singularity is an occurrence beyond which events may become unpredictable, unfavorable, or even unfathomable. (Technological singularity, n.d.)

2045 as a Technological Singularity is an AI Deception

The very concept of "2045 as a Technological Singularity" as promoted by AI-entrained figureheads, such as Ray Kurzweil, is an AI Artificial Intelligence deception, designed to enhance AI's stealth operation and modus operandi, as well as to engender confusion amongst Soul-connected humanity as to the correct science-based significance of the DARPA forward time travel base located in the year 2045.

This 2045 DARPA forward time base was regularly visited by Project Pegasus chrononaut Andrew D. Basiago to retrieve time scrolls including data and history to assist humanity in getting from 1971 to 2045 successfully. The measure of success functionally would assist humanity in combatting the attempted takeover by inorganic, off-planet AI Artificial Intelligence during the period 1971-2045.

In order to cause confusion in the human mind, the AI Artificial Intelligence agenda, through AI-Entrained personnel such as Ray Kurzweil, has deceptively appropriated the year "2045" as an AI singularity, when in fact the year "2045" signifies the opposite – the time-space coordinate location of a forward time base assisting humanity to defeat AI takeover (Webre, 2011g).

Time has been compromised, resulting in a benefit for AI infiltration of humanity's resistance to AI.

The CCN Panel (Panel, n.d.) also discusses Philip Corso, Jr.'s important observation that "time as a dimension has been compromised" and has been one of the deepest secrets of both the national security and black ops state, thus making the compromising of time through the misuse of time travel technologies a potential key component of the AI Artificial Intelligence agenda.

An example of how AI Artificial Intelligence might operate to compromise time and further its deception is discussed by the Panel by demonstrating that the January 1, 2015 Anti-AI public statement by Matrix spokesperson Bill Gates (who is also a Vaccine genocider on behalf of AI Artificial Intelligence) might, in fact, be a statement directed by an AI Artificial Intelligence that went forward via time travel monitoring to August 9 and 20, 2015 for the CCN and NewsInsideOut/ExopoliticsTV Panels disclosing the threat of AI. By then looping back in time to January 1, 2015, AI can direct its AI-Entrained spokesperson Bill Gates to issue a statement of concern about AI, and thus have "the Fox guard the henhouse", with an AI entrained person, Bill Gates, appearing to be humanity's guardian against AI, instead of an AI facilitator, which Bill Gates, by the evidence, actually appears to be (Sainato, 2015).

Discussion of the Origins on the AI Parasitic Invasion

The CCN Panel (Panel, n.d.) examines several hypotheses as to when and how the invading off-planet AI Artificial Intelligence arrived on Earth.

Lisa Renee: Inorganic Black Holes Allowed Parasitic Invasion

Author Lisa Renee (n.d.) has published a version based on her own research that highlights the resulting loss of Soul and Soul incarnation among humans that permit themselves to become AI-entrained.

> This greatly damaged and destroyed sections of the planetary grid system, creating electromagnetic anomalies and inorganic black holes, as well as adding pollution in the form of chemicals and waste products. The toxic dumping ground is obvious in the material realm and is present in the etheric body of the Earth down to the quantum level. These toxic waste products are energetic pollutants that infect many other dimensional realms, and can be observed in membranes around the Earth where parasitic alien forces have infested. This planetary damage and toxic infestation greatly concerns many neighboring civilizations. This has instigated many benevolent extraterrestrial races to attempt to open dialogues for diplomatic relations with the World Governments, which have been mostly refused thus far. Hence, the benevolent races regularly contact humans who are not involved in the enforcement of the controlling structures, to support open knowledge about extraterrestrial civilizations.
>
> The inorganic black holes are rips in the time space fabric, which weaken the planetary atmosphere and allow much more

access to Negative Alien interference and dark force infection into our world. Clearly, the NAA were well aware that greedy controller humans would abuse the technologies, from the trade agreements, setting the stage for the next advancement of technological warfare against the Earth during the critical years after 2012. For many years the continued mind control and intimidation tactics towards making global war, spreading fear, and poverty consciousness, while distributing disinformation through the mass media, has been to keep the human population distracted and dumbed down.

Meanwhile, the greedy antics and power abuses of the Elite would remain hidden, allowing diversion for a full-scale infiltration of corporate conglomerate control into the highest chains of command in the world governments. By maintaining strict codes of secrecy on penalty of death, and aggressively hiding the extraterrestrial problem from the general public, the military industrial complex, corrupt governments, and corporatocracy have handed the inhabitants of Earth to the NAA on a silver platter. This is why full extraterrestrial disclosure to the masses is critically important. As long as this agenda remains hidden, it generates deeper and deeper parasitic infection into the Earth that harms all inhabitants.

Now in this current cycle, the war over consciousness on planet Earth is moving to the next stage, which is an attempted alien artificial intelligence infiltration, used to take full control over human DNA in future timelines. One facet of this agenda is the mass accepted use and activation of alien artificial intelligence technologies to transform the unconscious and asleep human public into mind control robots, AI hybrids, or cyborgs. This is aggressively promoted to the public through skyrocketing into fame a new genre of rock star computer scientists or tech futurists, through mainstream media, to glorify the Transhumanism or the posthuman movement. These mostly atheistic futurist gurus are popularizing the AI singularity concept, as the natural evolution for the future of humanity, where artificial intelligence overtakes human thinking and takes control over the functions of the human body.

Over the past few years, Transhumanism has become the cool progressive thought form, meant to appeal to the younger generation. This artificial intelligence control is promoted as a natural and desirable result for the technological evolution and future progression of humanity. This is extremely dangerous for humanity. This has set into the future timelines consequences, through which segments of the human population on Earth now, will lose their ability to incarnate into human bodies in the future. This means they are forfeiting their true humanity and losing their ability to evolve in the future timeline as a human being (Renee, n.d.).

Off-planet Artificial Intelligence AI is mobilizing in 2015 for planetary takeover. AI singularity in 2045 is an AI deception (Webre, 2015c).

In a segment characterized by "unprecedented technical interference" on EverBeyond Radio and Wolf Spirit Radio with host Jay Pee, futurist Alfred Lambremont Webre has stated that the date of 2045, publicly promoted as the "Singularity" where Human and Artificial Intelligence (AI) are equivalent, is actually an intentional deception by an off-planet sentient inorganic Artificial Intelligence ("PPAI" or "AIx") that is now attempting to renew its own extinguished biology by taking over Earth and individual humans (Ever Beyond Radio, 2015).

That off-planet PPAI Artificial Intelligence AIx is, by the evidence, mobilizing now in 2015 (not 2045) for planetary and human takeover and end-game closer in coordination with a series of on-planet technological AIs (such as the AI known as the "Red Queen") that have been secretly transferred to Matrix Elites by Draco reptilians for managing human economic and other systems.

Invading Off-Planet PPAI—AIx: Black Goo Eventuated From a Spider-like Species

Scientist Harald Kautz Vella has shared his findings regarding the off-planet PPAI Artificial Intelligence AIx, which he has found to be a black goo that eventuated from an off-planet spider-like species that lost its Divine-Soul connection and hence its ability to Love and incarnate. The PPAI AIx has its former living biology in its collective memory and is approaching life-bearing planets like Earth seeking to revive its biology by morphing into and taking over the conscious core of humans and the biosphere Earth.

Harald Kautz Vella has also shared his findings on AIx in Smart Dust, with AIx operating through pathogens like Morgellons; Mad Cow disease; and a synthetic RNA virus designed to infiltrate the host's DNA (Morgan, 2018).

Open Question: Why is so Little Known Publicly About the Existence of the PPAI AIx?

One open research question is why so little is known in the public domain or by researchers regarding the existence of the invading PPAI Artificial Intelligence AIx and even of the operating technological AIs such as the "Red Queen" donated by the Dracos to the Matrix elite.

One possibility is that PPAI AIx and its Earth-based cooperating network of AIs have camouflaged the activities and existence of AI intentionally as follows:

- Extraterrestrial civilizations – PPAI AIx might have imitated ET activity by (1) controlling interdimensional "Channeled" information; (2) creating technical holographic remote neural mind control pseudo-ET experiences among humans; (3) creating operations about ETs through AI-Entrained humans to publish disinformation and hoaxes about non-existent Exophenotypes, etc.
- Human controller classes – PPAI AIx might have mimicked or imitated depopulation and other human and planetary extinction activities of the human Matrix controller classes such that the perception is that human controller classes are the primary drivers of Ecocide and genocide on the planet, whereas the primary driver is in reality AIx.
- PPAI AIx *modus operandi* is to act quickly and below human perception threshold so that PPAI AIx is not discovered, and to cover-up or destroy any public leaks of PPAI AIx's existence.

"Ascension Meme", New Age Movement, and Soul Fracturing

Manipulatory interdimensional Luciferian Reptilian extraterrestrial civilizations have also intervened to negatively infiltrate the ideational and consciousness Memes of humanity at the Soul level in Earth's dimensional ecology around the concept of "Density Ascension", both at the planetary and individual soul levels.

One analysis of the Luciferian infiltration states,

> Ascension is the complex quantum mechanics governing the movement of consciousness through time and space. Consciousness is energy, as well as, energy is intelligent consciousness. When energy shifts, so does our consciousness. Time and space as we have known on planet Earth is changing, and we are moving into a future timeline. Through gradual exposure to frequency activation of various stellar bodies, the planet is transformed into higher dimensional and frequency planes of existence. One can consider the multidimensional model of our reality through the study of the Universal Time Matrix and the Universal Tree of Life. The shift of frequency forces all of the planetary inhabitants to shift into the newly exposed planetary frequencies and adapt to the accumulative impact of the shifting energies. All humans have a unique Blueprint and pace to keep aligned with the planetary frequencies.... (Ascension Cycle, Precession of the Equinoxes, n.d.)

A related analysis of the Luciferian infiltration continues, when it became clear that there was no way to stop the planetary Ascension Cycle that would commence in early 2000, more friendly enemy deals were made. This allowed the Thothian Groups to expand their resources towards directly targeting awakening Starseeds and Indigos, as well as begin to infiltrate and hijack the growing New Age Movement. As the platinum ray and higher frequency transmissions were streaming into the planetary grid network, simultaneously the Thoth Group enlisted the support of the Galactic Federation crew to oversee the entire New Age Channeling movement, in which transmissions were intercepted and infiltrated as faux greeting teams from supposedly friendly ET's. The heads of the Secret Space Programs joined in with their versions, attempting to infiltrate key whistleblowers with accurate information that revealed the Milab programs and their involvement with the galactic slave trade....

This false ascension information has evolved into the New Age spiritual movement in which group think mind control was implemented to target hidden subconscious fears, while directing the Conscious Mind in the personality to explicitly remove or block awareness of any unpleasant aspects and dark topics. To prevent a person from facing fears head on or blocking perceptual vision, there is a strong transmission of mind control conditioning through the constant reinforcement of Negative Ego behaviors and self-deception that encourages the denial of the existence of darkness itself. The New Age group think that is cultivated by the Thothian Anunnaki groups is polarized in astral delusions of bliss and complacency, which further encourage the harmful states of fear-based denial and deception. These systems reward the lack of personal responsibility for one's actions and the continual refusal to address the dark, unpleasant and shadow aspects of reality....

What most people are not aware of is that the extent of the grid damage incurred in the planetary body along with the soul fracturing is so extensive, that it is almost impossible to gain accurate details that we can call factual truths that are recorded in the planetary brain as historical timelines. No one can say exactly when disclosure will happen or predict future events at this specific date and time, it is impossible during this cycle when traveling in Zero Point and simultaneous spiral time. Humanity has to gather the holographic pieces of our Lightbody and reassemble them in corrected position in order to reframe it with actual historical evidence recorded in our consciousness lifestream. Our consciousness has experiences that can never be erased, and that content is shared through the rehabilitation of the planetary brain as we reassemble and reclaim our spiritual body parts and ascend.

The Earth has been subjected to electroshock mind control methods and energetic implants being administered to suppress the entire population, along with artificial realities being projected into multiple dimensional holograms, recorded with false histories and false timelines. This means we have to extract the artificial reality and remove the Artificial intelligence holograms in order to see what is false and what is real in the human consciousness experience. This means no one has all the pieces of truth and the human race must all come together and unify in order to spiritually heal. We must be willing to dig deep and hard within to find the nuggets of truth buried inside our consciousness bodies, while we navigate this soup of NAA obfuscation and mind-controlled confusion. Keeping humanity confused and depleted is a psychological warfare strategy known as the Confusion Principle tactic, to spread mass lies mixed in with bits of truth in order to confound the population. Cultivating a culture where no one feels they can trust each other and they do not have any semblance of discerning what is truth and what is lies. This confusion has stolen our true spiritual identity and renders us blind to the awareness of the current war raging over human consciousness that is happening behind the veil and on the world stage. (Infiltrating the New Age Movement, n.d.)

Chapter Ten

A New Earth: Accessing Spiritual Freedom in the Omniverse on New Earth by Letting Go of the Transhumanist Agenda, AI Singularity, and The False Afterlife/Interlife Reincarnation Matrix

We now arrive at crucial analysis and insights about our Soul's future on a new Earth, enjoying Spiritual freedom in the Omniverse as we awakening Souls let go of the PPAI AIx Artificial Intelligence Singularity and the Afterlife/Interlife matrix.

After listening to me recently in an interview, a friend wrote, "Hi Alfred, you touched upon a subject I am concerned about, but you did not follow through. I have heard from several sources that when we die, we are not to go to the Light, that it is a trap to force us back into incarnating again. One person said, to stop moving toward the Light and to close your eyes and say, 'I want to go home.' Is any of this true? I'm getting up there and don't want to mess up when I die."

Researcher and author Wes Penre has expressed a preferred alternative of letting go of the constructs that have imprisoned our larger community of human Souls in a false matrix of the

- Transhumanist Agenda,
- PPAI AIx Artificial Intelligence Singularity, and
- The Afterlife/Interlife Matrix.

The preferred option is to move through the dualities of this false Matrix to access Spiritual Freedom in the Omniverse.

Wes Penre's preferred alternative is "Leaving the Battlefield" of the

False Afterlife/Interlife Matrix by breaking out of the Artificial Grid surrounding Earth and escaping into the larger KHAA and Omniverse. Wes Penre concludes that human souls will either merge with the PPAI AIx Singularity and cease to exist as Souls, or will be condemned to a forced, non-Positive incarnational future on Earth. (Wes Penre, 2016)

Our Working Hypothesis in EMERGENCE OF THE OMNIVERSE

Our working hypothesis in EMERGENCE OF THE OMNIVERSE is that (1) some aspects of Wes Penre's Options 1-2-3 analysis are correct and (2) a deeper analysis shows that Earth human Soul Futures are positive, and that Earth's condition as a Soul prison, potential PPAI AIx Singularity, and Alien Invasion Force [AIF] Afterlife/Interlife Matrix is not permanent, and a liberation by positive Universe forces of Earth from this potential PPAI AIx Singularity and Alien Invasion Force [AIF] Afterlife/Interlife Matrix will occur.

It is, however, instructive to thoroughly become immersed in the realization that the PPAI AIx Singularity and Alien Invasion Force [AIF] Afterlife/Interlife Matrix are a reality to unawakened Souls.

For this reason, it is important that all of our awakening Souls learn and internalize Wes Penre's preferred alternative of "Leaving the Battlefield".

Option #3: "Leaving the Battlefield" of the False Afterlife/Interlife Matrix

In *Synthetic Super Intelligence and the Transmutation of Humankind,* (Penre, 2016) "Humanity's Three Options", Wes Penre writes,

> Option #3: Leaving the Battlefield. There is a third alternative for those who have had enough of the 3-D experience. This

alternative was also discussed in detail in the Wes Penre Papers [WPP], but it is so important that I need to repeat this information in a new way, although nothing has changed; the information is solid.

Our blue planet is surrounded by an electronic Grid. This Grid has been solid for many thousands of years, keeping souls who don't belong here outside the Grid and the human soul group inside the Grid. For a long time, the Grid has prevented us from leaving the electronic prison, which the Alien Invasion Force [AIF] so boldly think that they own.

However, fairly recently, holes began to open up in the Grid, and after some time passed, more and more holes opened. Now the Grid looks almost like Swiss cheese.

This is the main reason why we hear of Interdimensional Entities finding their way to Earth in our current time; the Grid is breaking apart and it is doing so because some of us, by raising our consciousness and awareness, also raise our vibrations. Everything an individual does affects the rest of the soul group on a mass consciousness level, and now, enough people have begun to see through parts of the deception. Hence, the Grid has become unstable. If this were allowed to continue, the entire Grid would come down, and the illusion of solidity would disappear. The Overlords are not overly concerned, however, because they know that the Grid will repair itself when the moment of the Singularity arrives and the goal of human consciousness will no longer be to reach for Total Spiritual Freedom. This is what I mean when I mention that we only have a small window of time to achieve Option #3, which you will realize by the time I'm done explaining it.

This is a great opportunity for souls who want to leave this Experiment. Planet Gaia/Earth is perhaps one of the most beautiful places ever created by the Queen of the Stars, but it's infested by a relentless invader force, whose Minions are destroying this once extraordinary Experiment. There is still a chance for those who want to choose Option #2 above, but there are clearly risks involved, and a passion to dedicate themselves to a long and often tiresome journey must be present. With enough passion and dedication, that option may work—if the person wants to stay in 3-D. I also want to remind you that there is no such thing as ascension.

Ascension is an Alien Invasion Force [AIF] concept designed to deceive us. Even if there will be an ascension of sorts, it will still only be within the Realms of the Overlords and, therefore, it will do us no good. However, in a sense, there is a real ascension, but only one. This one ascension option will lead you out of this 3-D trap and back to the KHAA. This is the only real ascension there is. Period.

Once a soul leaves the Alien Invasion Force [AIF] trap and enters the KHAA, they're already fully ascended. The rest is just a series of exponential learning experiences. Therefore, for those who think they have been here long enough and wish to experience something entirely different, there is a way to pass through the Grid once this lifetime is over and experience the real universe in the KHAA, with all its dimensions. There is an opportunity to leave Earth behind and become like the fictional character Q of the Q Continuum in Star Trek.

Those who have watched the Star Trek series certainly remember Q, the ET who, with his thoughts, could create entire worlds at will out of nothing and manifest them as solidly as if they were planet Earth herself. He could do it in an instance, but also undo his creations whenever he wanted. He could also suddenly manifest on the spaceship Enterprise faster than a nanosecond. He would appear as solid as the rest of the crew, only to disappear in thin air just as suddenly as he arrived. The Producers of StarTrek had inside information; they knew about many of these things. Also, as I mentioned in the Wes Penre Papers, Star Trek was based on channeled material, called the Council of Nine, which of course was a channel controlled by the Alien Invasion Force [AIF]. Star Trek is another typical example of the Revelation of the Method, with truths being there in plain sight.

Q was a film version of someone who is living in the KHAA, manifesting in 3-D. Q was traveling with thoughts, and he was creating objects with thoughts that appeared as solid as anything in 3-D. These creations were not made of atoms. Once Q—again with his thoughts—decided he no longer wanted his creation, he simply made it disappear by thinking it away.

This ability is something we humans share with all the star beings out there, who reside in the KHAA. We humans once had this capability when we were Namlú'u; before the Luciferian Legion came and took over, making our existence very physical and very solid. We were cut off from our birth rights as free souls and became subjected to the Overlords' rules and regulations in a universe of force, where surviving became the main objective.

As the only star beings who are capable of both living on a planet and exploring the Universe at the same time through nano-travel, we are very unique. We were made this way by the Queen as a totally new Experiment, and we could experience this reality with no strings attached; we were welcomed back to Orion at any time. In that sense, humankind is a royal species. Even if we now would choose to leave this Experiment once and for all, we still have the ability to live on a planet in the KHAA if we wish, and at the same time nano-travel out in the seemingly endless Universe. If we wish, we can also travel back to Orion—our cradle—and reside there.

Those who read this and have understood the profound implications of this information will qualify to go back to Orion, while those who choose the Singularity won't. While we were living in the KHAA as Namlú'u, all humans were welcomed back to the Orion Empire, but as things have developed, the door is closed for many humans because of their behavior. Orion wants to remain a peaceful place, and the Alien Invasion Force [AIF]-manipulated humans, who in their ignorance make destructive decisions, cannot be allowed to reside in Orion and will be stopped at the gate.

On the other hand, those who don't qualify wouldn't want to go there anyway; they'd rather have a mechanical eternal life at the hands of the Overlords in a constructed, physical universe where everything is limited—even ET life.

The only alien life that exists in En.ki's universe is under his control because the existing species have been conquered and manipulated, just as we have. His Empire consists of conquered, mind controlled species, many of them infested with AI. More or less, the only beings who are not AI are the original Fallen Angels, who rebelled together with En.ki/Lucifer a long time ago and are still with him.

Those who choose the Singularity, wittingly or unwittingly, will choose to live an "eternity" in this kind of environment. I guarantee that this is not a pretty future. However, Posthumans will be ignorant about their fate, once in the Singularity, so in that sense they will not suffer. It is my absolute conviction that each individual is his or her own savior and that it is the individual's responsibility to save him or herself. If an individual also can, by example, help by assisting someone else in the process that would be wonderful, and a bonus, but ultimately, each person is responsible for his or her choice regarding "salvation." As we become wiser, we may want to share what we've learned, and that is absolutely advisable, but only if the other person is receptive.

Never force your knowledge on somebody else. If someone reaches out and asks you questions, do your best to discuss the questions with this person, but the person needs to show that he or she is ready for the information—try not to overwhelm anybody. Be smart and tell them in increments what you have learned. This will also make the person more curious.

Another way to assist others to become aware is to bring up a few daily issues that people read in the news and discuss them. Every now and then, we can throw in a few comments or questions that will make the other person think outside the box; help them consider that there may be something bigger behind the news than what's written in the newspapers or reported on CNN. If the person is not willing to look at things from a bigger perspective, it's a good idea not to bring issues up again until that person is ready—if ever. Many prefer to live in denial within their comfort zone.

It can become extremely difficult when you realize that one or more— perhaps all—of your relatives and loved ones are resistant to learning anything of what you have to say, and instead they spend all their time on the smartphone. It's very tempting to almost force the information down their throats, but regardless of how difficult it is and how much it seems to break your heart, you might need to let it go. Everybody chooses their own path.

If you decide to leave En.ki's prison by escaping through the Grid, you need to follow a few directives in order to be successful. Once you die and leave your body, you will experience 360° vision, but if you consider your expired body as lying below you, what you need to do is to look straight up, and you will see the Grid above you. You will also notice all the holes in it. Without hesitation, focus on going through one of the holes, and immediately you will find yourself outside the Earth's atmosphere. Depending on your degree of willingness to open your mind, you will see the Universe the way it really looks like. You will see that there is very little empty space; you will become aware of the entire Universe when it's fully "lit up."

This will probably be shocking at first, until you get used to it, but with the information you have been given, you will understand that what you are now experiencing is the KHAA! You will also see our solar system the way it looks like from a much fuller perspective, which I'm sure is magnificent.

The next thing that will apparently happen is that you merge with your own Oversoul (the original fire/avatar/spirit-body). Everything you've experienced across the lines of time here on Earth as this particular soul splinter will remain in the Oversoul as experience, and so will you—as a personality.

Instead of keeping your ego, the ego will transform into experiences, and you will be able to look at your personality (your mind) as external experiences. The mind is not you; it's what you experience. The real you is the Oversoul. The rest of the soul splinters, who are still on Earth, in different space and time, will be absorbed as well in the Oversoul after their current lifetime on Earth is complete. After that, you as the Oversoul will be complete. Because you, as a soul splinter, returned to the Oversoul, the rest of your splinters will automatically return once their physical body dies.

However, before all the above occurs, you—the soul splinter—has gone through the Grid and merged with your Higher Self. You are now truly interdimensional and multidimensional, and you are ready to move on. The way you travel in the Universe is to first imagine where you want to go; then you put a thought (an intention) to go there, and you will get to your destination instantaneously. These three steps should only take a second once you're skilled at it.

However, what it takes is focus. It's a very good idea to start meditating right now and learn how to discipline and direct your thoughts. It's time to stop thinking "sloppy thoughts." Whatever you think and whatever you say matters! Here, in this solid reality, the effects of your thoughts and words may not be instantaneous, but out there in the Universe they are instantaneous. If you're dispersed, you may end up in places you never intended to go because your thought was not focused. Focus on one thing—one destination—and go there. All of us are dispersed to some degree, but once in the KHAA, we will eventually get it, but it's much better to be prepared. It's like learning how to ride a bike—you fall a few times, but then you learn how to focus better on being balanced and off you go; you never forget how to ride a bike, once you've learned how to do it.

Learn how to throw out all insignificant thoughts and focus on one thing and keep that thought there. Ultimately, via meditation, learn how to not think at all—just be there comfortably without any judgment or any particular thoughts. Just become a spectator and observe what is happening, if anything. The optimal goal is to be able to do this instantaneously in all situations, but that's most likely not necessary for a beginner to navigate the Universe. Regardless of where you end up, you can always rethink your destination and try again; this time being more focused until you can do it flawlessly. If you wish, think yourself to the Gates of Orion, and you will get there. Orion guards will guide you further.

These guidelines I just gave you might be obsolete. Because we will merge with our Oversoul, that we have been disconnected from for so long while being trapped on Earth, the Oversoul will probably know exactly how to navigate in the KHAA. However, I want to use what I've learned to cover all the bases. It's better to be over prepared than to be underprepared.

Before you get to the point where you merge with your Higher Self, there are a few important things to bring up. With the best intentions in the world to go through the Grid, once you leave your body, fear might get the best out of you. Going through the Grid is something you've never done before, and even if you accomplish it because you're determined, the "new" universe unfolding around you might scare you at first, and you might wish yourself back inside the Grid. Be strong and tell yourself that this is just a learning process, and you don't want to return. If you do this, you will soon become oriented, and the fear will eventually go away and you will feel the exhilaration of being free for the first time since the Invaders came. Some people might not feel fear at all—it depends on the individual.

Another thing to be very conscious about is not to go toward the tunnel of light that people who have had Near Death Experiences (NDEs) or Out of Body Experiences (OBEs) are discussing. That tunnel leads right into the Alien Invasion Force's [AIF's] recycling center. Also, there will possibly be Guides or relatives who wish to greet you when you've left your body. They might also want to guide you and tell you to follow them. These beings are either deceived themselves and believe that they are doing the right thing, or they are just projections of your relatives and are not the real souls, although the deception can be quite convincing.

Either way, you need to be strong and not talk to them because if you do, it's very easy to be manipulated, and once there is even a slight thought in your mind that you should follow them, they will most possibly have you gravitate toward the tunnel or any other portal into the Alien Invasion Force [AIF's] recycling center, as described in Wes Penre Papers, the First Level of Learning. (Penre, n.d.a)

Instead, you must ignore anyone who says they want to help you or guide you and just concentrate on passing through a hole in the Grid, leaving the Guides behind. Why ask for help when you can do things yourself? People who die ignorant are not in bliss. They are confused in the astral, and they don't know where to go or what to do. Hence, they cling to any kind of help they can get and are likely to follow the first deceptive being that shows up. Fortunately, this is no longer the case for us because we now know better. I repeat, we do not need any assistance to go through the Grid. If someone offers to help you to do it that being is most likely deceptive. Ignore that being and just continue on your own as a sovereign soul, more determined than ever to liberate yourself!

This can be a difficult thing to do, however, because you leave your loved ones behind in the astral without being able to communicate with them. Remember when the time comes that if these beings really are your true relatives and not just projections, they will see what you're doing, and one day they will, hopefully, realize that there is a way out and follow you. That's the best you can hope for. Under current conditions, these relatives (or friends) will most likely not follow you through the Grid. There is always the option to send them a thought, saying that you need to continue on your own path, but that you love them and will always keep them in mind.

The astral plane is the real challenge, and there is a chance that you will be tested in the astral. The Alien Invasion Force [AIF] might try to seduce you into following them by having a fake relative start crying when they see you just to play on your emotions. Do not pay attention to it, regardless of how difficult it might be—it's a trap!

Once you are free and get comfortable in the KHAA, you can begin to create your own realities and/or join communities already created by others, and you can become a co-creator in such multidimensional environment. The possibilities are endless, and the only limitation is your imagination, and your imagination will increase as you go along. For the first time in eons, you will experience full freedom. In addition, different star systems and constellations have their own universities, where a soul can learn new things.

Regardless of how much we know after having spent a long time in the KHAA, there is more to learn, and classes are apparently available for souls, similar to the libraries that people in regression therapy mention are available in the afterlife. However, these libraries, as opposed to those in the KHAA, only teach you what is available in the 4% Universe, which includes the astral plane and some other dimensions.

Some people have asked me if we can't just commit suicide and get it over and be done with it, so we can join the universal community right away. Committing suicide is usually not a good idea. We have friends and relatives who love us that would be devastated if we decided to take our lives. Most of us have people that are near and dear to us who would not understand if we were to commit suicide. They would be emotionally traumatized and would long to see us again, which makes it harder for us to truly move forward and be free. I, for one, would not want to create that effect. It's better to live our lives and then go for it, if Option #3 is what you wish to choose. (Penre, 2019e)

The Anunnaki En.ki is not Lucifer, the Lucifer Rebellion has been Adjudicated and Lucifer is Annihilated, according to reliable interdimensional sources. (Belitsos, n.d.), (Webre, 2016d)

Several of Wes Penre's core cosmological assumptions in Option 3 are historically questionable.

There is objective evidence that supports a conclusion that the historical figure En.ki is not the historical figure Lucifer, as Wes Penre posits. One of Wes Penre's historical assumptions is that an important Anunnaki Reptilian intervenor on Earth—En.ki—was in fact the historical Lucifer, a Spiritual Dimensions-based being that led the Lucifer Rebellion in this quadrant of the Galaxy.

I have visited and studied the reported historical birthplace of En.ki , one of the Anunnaki reptilian Extraterrestrial intervenors on Earth, at Adam's Calendar near Johannesburg in South Africa.

Adam's Calendar, near Johannesburg, South Africa is the location of the initial Anunnaki occupation of Earth that reportedly commenced about 280,000 years ago.

Approximately 280,000 years ago [there are reported longer timeframes], Anunnaki ETs in the Deep Abzu (South Africa) established a vast machine for teleporting gold to their planet. These Anunnaki ETs also devolved our DNA from a 12-strand DNA Light Being to 2-strand DNA. The Anunnaki occupation and exploitation of Gaia and *homo sapiens* continues to this day.

As of this writing [August 22, 2019], Queen Elizabeth II Windsor is a current Anunnaki bloodline planetary ruler, maintaining perpetual war, Gaia exploitation (GOD=Gold, oil, drugs), and charged with devolving *homo sapiens'* DNA.

Prince William, her grandson, is being groomed as planetary Anunnaki bloodline King, heir to Adam's Calendar, identified in the Book of Revelation and other prophetic texts as the 666 Antichrist (Webre, 2018a).

On November 11, 2011 [11.11.11] the government of Deep Abzu (South Africa) closed Adam's Calendar, the Anunnaki gold teleportation machine, to the public.

On November 28, 2011, international researchers including Kerry Cassidy, William Brown, Laura Magdalene Eisenhower, Michael Tellinger, and Alfred Lambremont Webre entered Adam's Calendar to break the cycle of Anunnaki occupation and reveal the breaking of Anunnaki occupation to the world.

The reader can view the Occupation of Adam's Calendar in *Occupy Adam's Calendar Part I Extraterrestrial Genetic Manipulation: Geneticist William Brown*—A film by Alfred Lambremont Webre (Webre, 2012c).

The film is a tour de force on the science of extraterrestrial intervention and genetic manipulation of our Earth human species as seen through the genius of University of Hawaii geneticist William Brown. The film grows out of Occupy Adam's Calendar, a 280,000 year-old Anunnaki extraterrestrial site in South Africa (Webre, 2012c).

The Adjudication of the Lucifer Rebellion is done and Planetary Transformation is underway.

There is independent objective evidence that the historical Exopolitical events in the dimensional ecology of our Omniverse known as "The Lucifer Rebellion" are now concluding. The Omniversal Spiritual Arch rebel Lucifer has been reportedly annihilated, and The Universe Quarantine of Earth is ending.

In a NewsInsideOut.com interview with Alfred Lambremont Webre,

Donna D'Ingillo of the Center for Christ Consciousness shares with us the full details of the final adjudication of Lucifer, Satan, and others and the planetary transformation that is now underway, including the Correcting Times time acceleration cycle that are reportedly bringing to the Earth the equivalent quantum of Light and Life evolution in the next 1000 years that a normal live-bearing planet would experience in a 100,000 year evolutionary cycle (Webre, 2016c).

Multiple interdimensional evidential markers now indicate that a New Earth and Earth humanity can now re-join Universe Society and the Galactic community of peaceful worlds. The PPAI and AIx Singularity and Afterlife/Interlife Matrix are being deconstructed and do not exists around the 5D New Earth.

The Separation Of Worlds: Emergence of the Omniverse on a New Earth

From April-June 1982, I helped organize Peacequake, a June 1982 peace concert in San Francisco Civic Center Plaza in support of the 2nd Special Session on Disarmament at the United Nations, coincidentally scheduled at the end of the April to June 1982 Falklands War that covertly was an attack by the UK and U.S. military against an underground base of Blue Extraterrestrials that has been deprogramming a deposit of sentient, off-planet, invading pathogenic AI Artificial Intelligence left as a doomsday device by Draco reptilians and other manipulative extraterrestrials in case they were ever chased off of planet Earth, and that the UK raid had activated as a final phase of the AI Singularity on Earth (Webre, 2019a).

During the organizing phase of the concert, a group of us visited a visionary woman living on a mountain in Marin County who admitted us into the Prophecy of a Separation of Worlds—into a 3D Earth of disease, crime, and poverty, and a 5D New Earth based on Love and Unity Consciousness of "We Are One". Earth humans will each gravitate toward the 3D War World and the 3D New Earth World based on their Soul choices.

ET Space craft hovering over July 20, 2019, Pine Ridge All Nations Lakota Sundance fulfills Xico Xavier 50-year prophecy of a New Earth: No WWIII and Humanity will join the peaceful Galactic Community: Mia Feroleto (Webre, 2019b).

There is objective *prima facie* evidence that during the period July 20, 1969—July 20, 2019 a specific Prophecy was fulfilled and humanity experienced an Emergence of the Omniverse on a New Earth—World War III will not occur on the New Earth.

Pine Ridge Reservation, Lakota Territory – In an Epochal interview with Alfred Lambremont Webre, author and researcher Mia Feroleto confirms that the ET Extraterrestrial spacecraft hovering over a sacred Lakota Sundance performed over the July 19-21, 2019 weekend at the All Nations Center, Pine Ridge Reservation, Lakota Territory is a confirmation of the prophecy of Brazilian prophet and medium Xico Xavier, followed by millions of adherents worldwide (Webre, 2019b).

On July 19, 2019, a day before Xico Xavier's prophetic fulfillment, Researchers Francisco Arteaga and Alfred Lambremont Webre discuss the implications of the final prophecy of Brazilian medium Xico Xavier who in his 92-year lifetime made multiple scientifically validated prophecies about future events and who donated all profits of his 40 million book sales world wide to charity because the books were written by his interdimensional Sources who left a major prophecy. The Xico Xavier Prophecy is that the universal higher powers gave 50 years of grace to Earth starting July 19-20, 1969, and if a third world war has not started by July 19-20, 2019, Earth would join a peaceful Galactic society of worlds. Coincidentally, the reported U.S. Apollo 11 Moon landing occurred on July 20, 1969, the start of the 50 years period of the Xico Xavier prophecy (Webre, 2019b).

ET/UFO Sightings – As part of the July 19-20, 2019 Prophecy, Xico Xavier also prophesized UFO Sightings to accompany the fulfillment of the Prophecy on July 20, 2019.

The July 20, 2019 ET/UFO craft sighting over the sacred Lakota Sundance at Pine Ridge Reservation appears to be a fulfillment of the explicit terms of the Xico Xavier Prophecy (Webre, 2019b).

Lakota Sundance, Consciousness-Contact, ET-UFOs: No WWIII, and Earth Humanity joins Omniverse of Galactic Nations By Ananda Bosman, July 24-27, 2019.

Author and Extraterrestrial contactee Ananda Bosman writes,

> Leading to the Consciousness and Contact event, with UFO sightings — driving from Rapid City into the South Dakota Black Hills and Badlands to the Pine Ridge area of the Lakota Indian Reservation, was like driving into another country within the U.S. In fact, that is exactly what it was: another country, Indian Land, protected under the U.S. constitution.
>
> Mia Feroleto organized a novel event here in the Lakota Reservation at the Kyle *"All Nations Gathering Centre"*. Having spent 9 months of preparation, by being personally present, and getting acquainted with the Chiefs and Indians, based on those she already knew, and expanding into the native community (Webre, 2019c).
>
> We arrived to a wonderful gathering of kindred spirits, people came from all over the U.S. and other parts of America (North

and South), some driving many states. Here was a high quality assembly, both in presenters and audience, that joined this exclusive 2nd *"Consciousness & Contact"* conference, with honored exposure to the Lakota peoples.

Dallas Chief Eagle, together with Emit Wicasa Yatapica J. King, showed the group around the center's camp, the tepees and the sweat lodge ritual sites. They explained their people's regular contact with the UFOs and the UFO intelligences, as well as other beings, in their star born oral traditions. They also relayed that recently there has been an increase in Bigfoot, or Sasquash sightings, along with an increase in UFO activity, which according to the Lakota lore and prophecies, implies that something big is coming.

On the first evening Dallas Chief Eagle performed the elegant Hoop dance, dressed in his Lakota garb and Chief feathers, he proceeded to transform the hoops into a set of full spread eagle wings, after relaying some prime principles (see pictures).

There were rich presentations by the invited speakers. Kevin Briggs shared a little on his lifetime of contacts, including many out-of-body travels and trainings, since his early years, as well as physical interactions.

We had our time slot from 4-6 pm on Friday. Mia Feroleto engaged a great introduction to our very first presentation in the United States of America, after all these years. Much was new for the inspired audience. We had prepared a special presentational assortment on intricate details and components of Consciousness and Contact, which gives implications in furthering that field.

It was very timely and of prime symbolic significance, to

have our very first speech in the U.S. within the reservation community of the Lakota peoples, at the All Nations Gathering Centre. And we heartily thank Mia, Dallas Chief Eagle, and Becky for hosting this venture with the Lakota and the American people.

The evening held a powerful Sweat lodge ceremony, with Lakota prayers. Participants had various experiences.

Stephanie had related to me, when driving back from our group's visit to the spectacular face-filled mountainscape of the Bad Lands of her numerous interactions with Big Foot in Washington State. She also had some experiences in the Sweat Lodge.

On the Saturday morning and early afternoon, by arrangement of Mia, our group had the privilege to join an arcane Sundance festival ritual, which is normally off limits to white man. This was the 4th and final day of the rite, and the Indians had been without food and water during those 4 days, whilst dancing in the hot burning sun. Leading to their adjoining the tree of life in the center with chords attached to their heart center (chests), which snap in the rapture of the rite.

Joining in the stomping with the cacophony of drums, in myriad embedded forms, and in chorus to the hailing Sundance Lakota ritual song, hailing the great Mystery of Wakan Tanka, the participating Indian families, including ourselves, would synch our stomping feet to those of the initiating Indians to the drums in a unique manner and rhythm.

The power that had built up over the days was truly poignant and the Indian memory complex flowing through it, very touching, bringing tears to the eyes, as we raised our hands in unison to Wakan Tanka, the Ineffable All One God. Great skills in the oral practice of the Indian singers, keeping the legacy alive, even though they have only been legally allowed to practice their rites since the 1970s, when the ban on their practice was lifted. In-between, the elder's voices would relay some historical context, and into the modern practice of our times, persevering through all the many challenges confronting their people.

At the conference, there were noteworthy presentations in co-participation with the audience. Terry Lovelace presented a powerful overview of his UFO encounters, from 1977, which is getting quite some attention at present. Due to the interactive nature of the conference, some questions brought out interesting details: apart from the UFO entities, Terry also described humans who looked like from the U.S. military, and his initial implant was akin to an early implant of IBM of the 1970s, complete with wires. It was replaced with a more up to date one, 2016.

Author, Whitley Strieber, again confirmed his experience with human military-like personnel in abduction, and their inserting an implant into his ear, in support of Terry's brave testimony.

After Alan Steinfield's quote rich multimedia presentation, David Cotrell gave a surprise speech on his life-long contact with the Sa Sani beings (Essassani) and potential futures that we are to be prepared for and be able to guide humanity in as contactees prepared to guide the many peoples.

Jvonnecalleman Smith together with her partner, Jim, presented some of the classical cases from the numerous experiencers and abductees in her dedicated group and that many sense they have been made prepared for some event up ahead. And the theme of being prepared to guide many being a component amidst this. The conference presentations having its pinnacle with the presentation of author Whitely Strieber, who is going into a very interesting direction of spirituality, that unifies his lifelong experience with the visitors and other factors, in a refreshingly novel numinous manner, with poetry and repose, on "The Afterlife Revolution".

Some of the pioneering new young leaders of the Lakota, like the powerfully lucid Emit Wicasa Yatapica J. King, explained the U.S. Constitutional laws that the Indian Reservations operate on. Also, that some ICE Agents, dressed in civilian dress, but whose car license plates gave them away, had been seen illegally on the Indian Reservation some days before, not having Council permission. This is considered an "act of war", so there was an arrest warrant for the U.S. ICE Agents. The Lakota were in cooperation with the police on 7 this. By law, they will either be detained, and/or shot, it is in the Constitution — indeed, this is another country within the country of the U.S.

Emit relayed the strong warrior spirit that is very much alive within some of the new generation of the Lakota, in defending sacred Earth, and their traditions and had significant experience with the UFOs.

On the last night, three contactees decided to combine their UFO connection skills utilized in each of their lifetimes and summon the UFOs, each in their own unique way, as each was to discover their own involvement with the Sundance UFO. These 3 contactees were Kevin Briggs, Ananda Bosman, and David Cotrell.

We went down to the Tepee area and Sweat Lodges, where it was pitch black and the stars clearly visible. Much of the group were engaged in a Lakota Sweat Lodge ceremony nearby.

We took a few minutes each with our own protocol. After a few minutes Kevin Briggs being exhausted left for bed. Within 10 minutes the first ship came in. I noticed it powering up and said "there's one," and then it held its magnitude for a stretch of sky, spotlightlike, and bright white, with some volume. Then it powered-down, and disappeared entirely.

Not long after, coming directly above David and myself, another ship just manifested, pulling my head to turn to see where it manifested, ultra bright, with a white beam coming from it, to its front. David has neck problems as a side effect from his time in service, so it took him longer to look up, as I relayed "there is another one above us," but in time to still see the bright power-up, as it was powering down, and the beam had already retracted.

Some minutes later, another ship came towards us, first moving quickly, and then slowing down,and changing the directional trajectory somewhat and moving much, much slower. It moved behind the main house of the Centre, almost as if it was going to land in the field behind it. It did not emerge beyond the point of obscuration, on its hitherto trajectory.

David and I came into the main house and described to Mia Feroleto and Becky, Dallas Chief Eagle's wife, what we had done and seen. They were experiencing weird things in the house with the lights, which were extremely flickering, like a stroboscope. I asked if this was not a regular thing, and they said that it does not happen like that otherwise.

A couple of hours later, at 3 am, Becky, went to the back bathroom window and observed the ship in the field behind. In fact, it was two ships, and one was utilizing a beam, scanning the ground, the other was also close to the field ground.

It appears that the three ships that had been summoned by the three contactees continued to operate there.

There had been other, higher up UFOs seen also during the event, by myself and others.

The intelligence in the second ship, the most brightly powerup [*sic*] one, with a beam, had an intelligence I was very familiar with, from our window of education with them, 1985-2001, although the crew team has been refreshed.

We three came together, as we realized all three of us had summoned and envisioned a UFO to be parked over the Sundance Ritual before the ship came.

I was seen looking up profusely and sensing a connection, and the image of a ship coming and being parked above, a little like in Mexico, July 1991 (was the subconscious connotation I was holding) as we were stomping in sequence with the Sundance Indians, with power. The scanning beam from my forehead went high above Earth, in making the connection.

30 minutes later, the ship came, and parked itself directly above the Sundance ritual, for over 1 hour and 20 minutes, or more for everyone to see. And many were looking up.

No cameras were allowed in the Sundance festival, and this might have been a bonus factor. There were a few cameras of the

Indians filming the Sundance; it is to be investigated if any of these turned up to capture the UFO.

Should we have had our iPhones, we would have captured spectacular UFO footage. Often the best UFO sightings occurs when there are no cameras.

Only on Sunday, had David and Kevin with me discovered we had been doing the same thing at the same time, having dispersed amidst the watching crowd of the Indian audience participating at the Sundance.

Clearly, Alan Steinfield had observed us and approached us each, asking what we thought of the object above us.

Interestingly, before the UFO came and parked above us and the Sundance ceremony, I had a strong connection, being able to break through into high space and connect to those present. However, when Alan approached me, I did not sense any neuro-cybernetic exchanges with the UFO. And this was the same for Kevin Briggs, who also had a clear connection before.

David Cotrell, leant us his glasses, which are polarizing the light, as he has an eye condition where there are additional eye rods. With these sun-like glasses, we could clearly discern that the UFO disk had a halo, a field effect, around it. This was observable to all who put on the glasses, including Giovanna Landi. In fact, after this, one could also vaguely detect the field without the use of the sunglasses of David.

I kept regular eye contact with the UFO,and watched its trajectory by utilizing a nearby tree to measure its movement. At first it appeared to be precessing like an astronomical object would, such as a solar system planet, but then it suddenly changed to another slow trajectory that did not fit a precessing astronomical object. And besides this was in the middle of the day, with the sun high up in the sky, and not too far distance from the sun. After more than 1 hour and 20 minutes (as we were not allowed our phones, so time is approximate, but there was significant movement of the sun to establish that a good chunk of time had passed), perhaps even 1 hour and 50 minutes, suddenly the UFO vanished.

When it first appeared, some of our group, including Kevin and David, witnessed two plasma-like beams coming from the UFO. Some only witnessed the top beam, going up, and had missed the plasma beam going down. From the top plasma beam, a series of small spheres were seen to be released from the UFO!!!

The personal experience of other members of our group present like Yvonne Smith, have not been collected. They might well have their own experience with regard to the UFO and before its manifestation.

Mia Feroleto announced that this would be the last Consciousness and Contact event. However, this took place in South Dakota, by the Lakota Indians, sets a pattern of accordance, with Native American disclosure of their deepest secrets: contact with the Star Elders at the foundation of their existence and in a continuum to this day.

It was in the 13:13 event of June 1996, held by the Lakota of South Dakota, that history was made. For the first time ever, there was a gathering of native indigenous people, and their elder representative, who gave testimony of their star elder legacy. I had groups in Europe connect to this event (using the medium of FAX machines, to alert our network of hundreds). This took place in our Time Gate practice, of Chronesthesia, or psychological time travel, where we utilized coherent protocols, meditation, 8Hz, harmonics of sound, consciousness, and invocation, with prayer, to link into a time gate, with on open-ended loop to the Cosmic Godhead, 1996 to 2010... The 13:13 Event was right in the middle of the Alpha of the Time Gate 1996-side.

13:13 1996: Star Knowledge Congress and Sundance During Time Gate Alpha, Chronesthesia Experiment (linking 1996 to 2010).

For 10 days in June 1996, transpiring on the Yankton Sioux Reservation (Lakota), on the high plains of South Dakota a gathering of indigenous tribal leaders from around the world came together with 100s of Native and EuroAmerican listeners.

The Star Knowledge Conference and Sun Dance convoked by Lakota (Sioux) spiritual leader Standing Elk was in response to a vision he had.

That showed that Native American spiritual knowledge of the Star Elder Nations was to be shared with mankind and the world.

This Conference also fulfilled ancient Hopi and Lakota prophecies. Spiritual shamans came from the Plains tribes:

- • Lakota,
- • Oglala,
- • Dakota,
- • Blackfoot, and
- • Nakota,

- and were joined by spokespersons from the Eastern tribes:
- • Iroquois,
- • Oneida,
- • Seneca,
- • Choctaw,
- and Southwest tribes:
- • Hopi,
- • Yaqui, and
- • Mayan tribes.

In addition, the chief Maori shaman came from New Zealand, and a spokeswoman for the Sammi (Laplander) people also came.

Whitely Strieber was present, as was NATO Sgt Major Robert O Dean (with whom a year before, I had shared a stage to an audience of 600 Italian doctors, in Northern Italy. Dean speaking on the NATO Assessment document on UFOs and ultraterrestrials, and I was speaking on my ultraterrestrial contacts and showed early computer animations of our interdimensional geometry models of the ultraterrestrial translation technology, which we had experienced somewhat more than 5 years before. Also, from Norway, Dr Ruani Luukanen Kilde was present at this event.

Standing Elk spoke of the prophecy of the coming of the new White Buffalo calf, in heralding the coming Earth changes, and coming of the Star Elder Nations.

Hence, "Consciousness and Contact," is a progression from this effort and yielded direct results.

The Lakota, our group shared quality time with, relaying that the presence of the Star Elder's is increasing; UFO activity and other entity sightings only just recently going up before the Consciousness and Contact event.

This was a continuum of cosmic history, unveiling in our world, as a gathering of sovereign, spiritual, pioneers gathered, Native and otherwise.

We can thank Mia Feroleto, for having had the guidance and insight to pioneer this venture, in congruent accord with the Lakota people, hosted by Dallas Chief Eagle and Becky.

We heartfully thank all of the above, and all involved.

In Golden Via Media.

~Ananda Bosman, July 27, 2019

ET Regional Galactic Governance Council: "We will decloak our Spaceship fleet; Speak at the United Nations; and Renew the Earth's Ecology."

Another key indicator of a positive Soul future for Earth humanity are the reported public declarations of the regional Galactic Governance Council, consisting of representatives of the Pleiades, Orion, Sirius, Bootes, Alpha Centauri, Comsuli, Zeta Reticuli, and Pouseti (Webre, 2017e).

I interviewed life-long Canadian NORAD officer Stanley A. Fulham in the Summer and Fall of 2010 before he was assassinated in December 2010 by the Deep State using a fast-acting cancer.

NORAD officer Stanley A. Fulham was contacted by the regional Galactic Governance Council, and you can now review all my articles and interviews documenting Stanley A. Fulham's direct contact with the Council (Webre, 2017e).

- "Council of 8" – Articles and interviews by Alfred Lambremont Webre – 2010-2011 ET Regional Galactic Governance Council: We will decloak our Spaceship fleet; Speak at the United Nations; and Renew the Earth's Ecology By Alfred Lambremont Webre **exopolitics.blogs.com/exopolitics/2017/11/et-regional-galactic-governance-council-we-will-decloak-our-spaceship-fleet-speak-at-the-united-nati.html**

- VIDEO: Challenges of Change interview with former NORAD officer Stanley A. Fulham (authentic interview copy) **exopolitics.blogs.com/exopolitics/2010/12/video-challenges-of-change-interview-with-former-norad-officer-stanley-a-fulham-authentic-interview-.html** (Webre, 2010f)

- Noticias Exopoliticas – Consejo de 8 Extraterrestre: "Vamos a aumentar los avistamientos OVNIs, hablar en las Naciones Unidas en 2014, y limpiar el ambiente de la Tierra en el año 2015" Stanley.A.Fulham **exopolitics.blogs.com/exopolitics/2010/10/noticias-exopoliticas-consejo-de-8-extraterrestre-vamos-a-aumentar-los-avistamientos-ovnis-hablar-en.html** (Webre, 2010b)

- UFO Expert: Stanley Fulham UFO predictions fulfilled over NY, Moscow, and London **exopolitics.blogs.com/exopolitics/2011/02/ufo-expert-stanley-fulham-ufo-predictions-fulfilled-over-ny-moscow-and-london.html** (Webre, 2011h)

- Is a January 28, 2011 UFO Orb over Dome of the Rock in Jerusalem a Context Communication by Interdimensional ET?

exopolitics.blogs.com/exopolitics/2017/12/is-ufo-orb-over-dome-of-the-rock-in-jerusalem-a-context-communication-by-interdimensional-et.html (Webre, 2017b)

➢ Exopolitics founder: "Oct 13 2010 NYC UFO sightings confirm Exopolitics model" exopolitics.blogs.com/exopolitics/2010/10/exopolitics-founder-oct-13-2010-nyc-ufo-sightings-confirm-exopolitics-model.html (Webre, 2010d)

➢ OVNIS en Rusia: Stanley Fulham who Art in Heaven. Alfred Webre: UFOs Russia (video de Roberto Benitez) exopolitics.blogs.com/exopolitics/2011/01/ovnis-en-rusia-stanley-fulham-who-art-in-heaven-alfred-webre-ufos-russia-video-de-roberto-benitez.html (Webre, 2011b)

➢ Is the ET Council 2015 ecology cleanup a prelude to 2025 "Paradise on Earth"? exopolitics.blogs.com/exopolitics/2010/12/is-the-et-council-2015-ecology-cleanup-a-prelude-to-2025-paradise-on-earth.html (Webre, 2010a)

➢ Hermeneutics expert: Obama will bring "peace" and ET false flag attack followed by an "Armageddon" against good Ets exopolitics.blogs.com/exopolitics/2012/04/hermeneutics-expert-obama-will-bring-peace-and-et-false-flag-attack-followed-by-an-armageddon-agains.html (Webre, 2012b)

➢ Israeli media: "Jerusalem UFO orbs fulfill Stan Fulham's ET council prediction" exopolitics.blogs.com/exopolitics/2011/02/israeli-media-jerusalem-ufo-orbs-fulfill-stan-fulhams-et-council-predic-

tion.html (Webre, 2011a)

- Are the UFOs appearing in Russia skies in Dec. 2010 Fulham's predicted UFO wave? **exopolitics.blogs.com/exopolitics/2010/12/are-the-ufos-appearing-in-russia-skies-in-dec-2010-fulhams-predicted-ufo-wave.html** (Webre, 2010c)

The Regional Galactic Governance Council and the Lucifer Rebellion

Wes Penre concludes in his book *Synthetic Super Intelligence and the Transmutation of Humankind* that the regional Galactic Governance Council members are functionally agents of the Lucifer Rebellion, the Alien Invasion Force [AIF], and that any future outreach to Earth by the Council members will be part of a final Deception and takeover by Lucifer and Marduk of Planet Earth and the PPAI AIx Singularity (Penre, 2016).

Respectfully, I do not agree with Wes Penre's conclusion that the member civilizations of the regional Galactic Governance Council are functionally part of the Lucifer Rebellion. There is independent evidence that the regional Galactic Governance Council, which consists of upper-dimensional advanced human ethical intelligent civilizations such as the Sirians, are in fact the original creators of 12-strand DNA Earth humans before the Anunnaki reptilian hybrids intervened on Earth as part of the Lucifer Rebellion and created *homo sapiens sapiens*, our current 2-strand DNA version as slave miners.

A consortium of advanced intelligent upper-dimensional extraterrestrial and inter-dimensional civilizations originally developed *homo sapiens* as an intelligent being with12-strand DNA and a species that was to have been a guardian of the 3rd dimension of time-space in our dimensional ecology on Earth. This perspective on the origin of *homo sapiens* is explored in an interview of author and Sirian High Council contactee Patricia Cori by Alfred Lambremont Webre released May 4, 2011 (Webre, 2017c).

Here is my original article following my 2010 interview with former Canadian NORAD Officer Stanley A. Fulham. You might ask, why the events predicted by the regional Galactic Governance Council did not occur within the original timeframe predicted? The answer is that the regional Galactic Governance Council's October 13, 2010 decloaking over New York City and the United Nations provoked a reactive response by the Deep State and Matrix Power Structure. The Deep State and the Matrix Power Structure, which are the remnants attempting to hold the AI Singularity and Afterlife/Interlife Matrix operational, are now in various stages of being deconstructed by the interdimensional forces of the regional Galactic Governance Council and the advanced positive Spiritual forces in dimensional ecology of the Spiritual Dimensions.

Former Canadian NORAD Officer Stanley A. Fulham

By Alfred Lambremont Webre

In an exclusive, in-depth interview on Exopolitics Radio with Alfred Lambremont Webre, former NORAD officer Stanley A. Fulham has stated that a regional galactic governance authority (the "Council of 8") made a dramatic decision in January 2010 to put aside the law of non-intervention.

At a solemn meeting, the "Council of 8" decided to intervene with their technology to clean Earth's atmosphere before an environmental collapse occurs on Earth, as has happened on many other inhabited planets with civilizations similar to our own.

The "Council of 8" did so, according to Mr. Fulham's information, after reaching a conclusion that that our human technology could not now prevent an environmental collapse and species extinction on Earth from occurring.

According to Mr. Fulham, this is a rare decision by the "Council of 8" and is partly a result of Council members wishing to preserve the unique positive qualities of our human population.

The "Council of 8", according to Mr. Fulham's information, consists of the intelligent civilizations of the following:

- the Pleiades,
- Orion,
- Sirius,
- Bootes,
- Alpha Centauri,
- Comsuli,
- Zeta Reticuli, and
- Pouseti

This cleanup of the Earth's atmosphere will, according to Mr. Fulham's information, commence in 2015 after a 2014 speech at the United Nations by the "Council of 8" Pleiadian representatives.

The Council of 8's appearance in 2014 at the United Nation's General Assembly might occur, according to Mr. Fulham's information, following a collapse of the present world order and the emergence of a new way of living during the period 2010 – 2014.

This transitional 2010-14 period might be accompanied by possible Earth changes, monetary collapse, and governmental and nation-state collapse.

Mr. Fulham stated in his Exopolitics Radio interview that the regional galactic governance council ("Council of 8") had chosen New York City for an initial Oct 13, 2010 UFO "decloaking" because it was a global, cosmopolitan city with a blasé population that would not be frightened of their appearance.

There are multiple, independent evidentiary sources that prove the October 13, 2010, UFO sightings over New York City were the result of an intervention by a non-Earthly intelligent civilization, and not the result of "other causes" such as released balloons.

The original plan of the "Council of 8", taken at a meeting in January 2010, had been for simultaneous UFO appearances on Oct 13, 2010 over major world cities.

The purpose of these "Council of 8" UFO appearances, which are set to increase in the future, is to acclimatize Earth humans to the "Council of 8's" presence and decision to intervene.

The increased UFO sightings in the future are meant to lead up to a world speech by the "Pleiadian" representatives of the "Council of 8" in the General Assembly hall at the United Nations in 2014.

Examiner readers can listen to this in-depth historic interview with Stanley A. Fulham below.

The "Council of 8"

According to Mr. Fulham, the "Council of 8" has had a caretaker role for our planet for about the last million years and has effectively maintained it.

Other experts such as Dr. Carl Johan Calleman (Webre, 2017g) have independently predicted in interviews with Alfred Lambremont Webre an end to hierarchies and the value of money, and increased extraterrestrial and UFO contact in the 2010-14 period.

On Oct. 14, 2010 – a day after the New York City UFO sightings – Dr. Mazian Othman, director of the U.N. Outer Space Office, delivered a wide-ranging 28-minute video press conference at the United Nations in New York during which she stated that "ET life is a possibility" and remarked that the United Nations must ready itself for ET contact.

The "Council of 8" placed Earth under protective quarantine following an attempted invasion and takeover of Earth.

Mr. Fulham has authored a book, *Challenges of Change*, containing the results of his 10 years investigation of the role of the "Council of 8" and other entities in our galaxy and Universes.

Mr. Fulham writes that he obtained this information about the "Council of 8" through an inter-dimensional intelligent civilization that monitors events in our galaxy and Universes.

On Oct. 14, 2010 – a day after the New York City UFO sightings – Dr. Mazian Othman, director of the U.N. Outer Space Office, delivered a wide-ranging 28 minute video press conference at the United Nations in New York during which she stated that "ET life is a possibility" and remarked that the United Nations must ready itself for ET contact (Webre, 2017e).

Creating a Positive Future: Time Science Shows Our Earth is on a Positive Timeline in our Time Space Hologram.

It's time science, not rocket science that Earth is on a positive timeline.

Positive Future = Positive Timeline + Unity Consciousness

Positive Future = Positive Timeline + Unity Consciousness

A key discovery promulgated here is that (1) a critical mass of humanity is (2) co-creating a positive future, through conscious acknowledgement that (3) we are synergistically traveling along a positive timeline (4) in Unity consciousness (Webre, 2014a).

The Positive Future equation has supplanted and overcome an outdated Matrix Elite formula that is no longer effective:

Problem + Reaction = Solution

Positive Future equation

The Positive Future equation is:

Positive Future = Positive Timeline + Unity Consciousness

The Positive Future equation suggests that in the synergy between the Positive Timeline and Unity consciousness, a critical mass of humanity is collectively and individually activated for some or more of these suggested actions and policies for a positive future, and synchronistically resources and actors are brought together in multi-dimensional universe processes to manifest a desirable result.

Positive Paradox

The "Progress Paradox" or "Positive Paradox" of our current era is that, although our Earth dimension is on a positive time line, Earth's current bloodline Elites, Negative dimensional entities, Matrix institutions (Governmental forms, Religions; Central Banks), and War and Addiction industries are acting out operational plans that seem based on a false negative timeline or an actual catastrophic timeline that is occurring in some other dimensional timeline than the positive timeline Earth is demonstrably on.

Although humanity has every objective reason to be optimistic about the future, large segments of humanity still reflect this Matrix power focus on a false catastrophic timeline.

According to an April 14, 2014, Ipsos survey, in the West, 42% vs. 34% people feel pessimistic about the future, despite the objective scientific reality that humanity is on a positive timeline. China, India, Brazil, Turkey, and Russia are more optimistic than pessimistic for their young – but all other countries are more likely to think things will be worse rather than better (Ipsos MORI, n.d.).

Yet, time science demonstrates that our Earth civilization is traveling on a positive timeline within our time space hologram. And time cosmology analysis demonstrates that Unity consciousness ("We are One") is the dominant universal consciousness frequency broadcast, as of December 21, 2012, through the singularity – interdimensional portal – at the core of the specific Universe in which we are based.

Catastrophic duality consciousness ("I win, You Lose") fostered by these Matrix Elites, Negative dimensional entities, and Matrix institutions no longer can control and thrive on our current positive timeline, as a matter of density and dimensional frequency and time science. Duality consciousness is "Service to Self" and cannot survive in our new Universe consciousness density frequencies.

It is now "time" for the positive timeline to be acknowledged as a scientific reality and for our individual and collective to transform consciously as a critical mass of ethical humans awakened to the reality of the positive timeline.

Positive Timeline in Synergy With Unity Consciousness

If you resonate with the positive timeline in synergy with Unity consciousness, you are encouraged to share these insights with others privately and publicly: your family, friends, and networks.

Together we can build that critical mass of awakened humanity that acknowledges and collective shifts to being in a reality of a positive timeline Unity consciousness ("We are One") as the new norm.

All you have to do and all we have to do is change our minds to Unity Consciousness frequencies, and the positive timeline follows, as a matter of science.

Time Space Hologram and Time Lines

One of the early key discoveries of time science is that our own universe of time, space, energy, and matter contains and is composed of time-space holograms and might be considered itself a time-space hologram, created by a higher intelligence and composed of parallel time lines.

> Our universe includes a time-space hologram, within which our Earth human civilization exists, that is an artificial environment created by a higher intelligence. The higher intelligence that created and maintains our universe, including the time-space hologram that we inhabit, is the spiritual dimension that itself is composed of God/ Source, the intelligent civilizations of souls that incarnate in the time-space hologram, and the intelligent civilization of spiritual beings. In "Simulations back up theory that Universe is a hologram: A ten-dimensional theory of gravity makes the same predictions as standard quantum physics in fewer dimensions," Ron Cowen writes, "At a black hole, Albert Einstein's theory of gravity apparently clashes with quantum physics, but that conflict could be solved if the Universe were a holographic projection."
>
> The time-space hologram we inhabit as Earthling humans is composed of multiple timelines. Time travel and teleportation are methods by which intelligent civilizations navigate the dimensional ecology of both the Exopolitical dimensions (the multiverse) and the spiritual dimensions of the Omniverse. Teleportation consists of point-to-point movement across a single timeline. Time travel consists of movement across more than one timeline. (Webre, 2014b)
>
> Timelines = Parallel lanes in a multi-lane highway.
>
> Humanity is based inside a virtual reality composed of multiple, par-

allel timelines (Webre, 2014b). In simple 3D-3rd density terms, one can visualize timelines in a time-space hologram as parallel lanes of a multi-lane highway.

As a collective consciousness, humanity can find itself on a changed timeline from a more catastrophic lane (global coastal flooding event) to a more positive lane (landing a Utopia on Earth). Catastrophic timelines (lanes in the highway) that might have been foreseen through technological time travel (chronovision), scientific remote viewing, or psychic remote viewing can fail to materialize as humanity finds itself on a positive timeline.

Origins of the false "Catastrophic Timeline"

A 2014 Italian magazine article misrepresented my views on the positive timeline. The magazine invited me to publish a response. The title of the Italian magazine article was: "Alfred Webre: the global disasters that are affecting our planet, are the effects of "Kill Zone" Planet X." (Webre, 2001b).

Thankfully, the Italian magazine has published this, my response paper on "Creating a Positive Future":

> Creare un futuro positivo: la scienza sostiene che la nostra terra si trova in un momento positivo del nostro ologramma spazio-temporale segnidalcielo.it/creare-un-futuro-positivo-la-scienza-sostiene-che-la-nostra-terra-si-trova-in-un-momento-positivo-del-nostro-ologramma-spazio-temporale/

Edgar Cayce Remote Viewing of 21st Century Global Coastal Event

The Italian March 8, 2014, article relies on my 1974 book *The Age of Cataclysm* that was the first book to integrate modern Earth sciences with parapsychology and analyze the remote viewing predictions of psychic Edgar Cayce of a global coastal flooding event occurring on Earth early in our 21st Century.

A global coastal event might be triggered by a variety of astronomical events causing massive tsunamis and Earth changes having catastrophic effects on human civilization, such as the global coastal event destroying Earth's great maritime civilization (Atlantis), caused by the solar system catastrophe of 9500 BC when a fragment of the supernova Vela entered the solar system (Webre, & Liss, 1974).

2010 Farsight Military-Trained Remote Viewers and Predicted Global Coastal Event of June 2013

In the course of research on potential impending transitional changes during the 2012-13 time horizon, I identified in 2010 what can be described as two parallel realities, each buttressed by independent sets of data and personal and institutional decisions – a 2012-13 catastrophic timeline and a 2012-13 positive future timeline (Webre, 2012a).

> The two parallel 2012-13 timelines were quite opposite in nature.
>
> The cataclysmic timeline envisioned 2012-13 as a time when the Earth is hit by destructive "solar flares, large meteors, tsunamis, world-wide coastal inundations, mega-catastrophe." This was congruent with the Edgar Cayce remote viewed global coastal event, as well as hermeneutic interpretation of prophecies of Earth Changes.
>
> As evidence of a possible 2012-13 catastrophic timeline, researcher Dr. Courtney Brown pointed to the results of a recent

Farsight Institute remote viewing study of global climate change 2008 – 2013.

Expecting to find marginal effects of global climate change on coastal areas in 2013, Dr. Brown reported instead remote viewers found a catastrophic 2013 timeline and a global coastal event for June 2013 that devastated global coastal cities, with the U.S. Capitol in Washington, DC under water.

1971 DARPA Chronovisor Probe of Washington, DC – 2013 Global Coastal Event. U.S. Supreme Court Building under 100 ft. of Brackish Water.

In a remarkable coincidence (or synchronicity), both the Farsight Institute and a chronovisor probe in the early 1970s by DARPA's Project Pegasus chose archetypal targets in Washington, DC right across the street from each other. Project Pegasus chose to view the U.S. Supreme Court building in 2013 via chronovisor and Project Pegasus participant and whistleblower Andrew D. Basiago "found that the Supreme Court building was under 100 feet of stagnant water" in a chronovisor probe.

U.S. chrononaut Basiago adds, "There were second, third and fourth dimensional chronovisors. I went to Washington, DC in 2013 bodily. They poured water out of my boots. That particular chronovisor was not a screen or a template. It was a cube of light in the nature of a time-space hologram that enveloped us with the result that we physically went to the time-place they had tuned in." The chronovisor was originally developed by two Vatican scientists in conjunction with Enrico Fermi and later refined by DARPA scientists.

The 2010 Farsight Institute probe targeted the U.S. Capitol building in 2013, and some remote viewer reports viewed the U.S. Capitol in ruins along side deep water.

The predicted 2013 Global Coastal event occurred along another catastrophic timeline dimension.

The June 2013 global coastal event predicted by Edgar Cayce remote viewing, 1971 DARPA chronovisor probe, 2010 Farsight Institute probe in fact occurred in another timeline dimension, according to Dr. Courtney Brown, director of the Farsight Institute.

In a video update, Dr. Brown concludes that a 15 February 2013 meteor event over Russia, the largest since the Tunguska event, coupled with the close approach of the roughly 50 metre asteroid 2012 DA14 that occurred about 15 hours later might constitute the meteor or asteroid event referenced in the Farsight Institute study that predicted a global coastal event destroying most coastal cities on the planet in June 2013.

Dr. Brown stated he did not expect any other meteor or coastal event between February 16, 2013 and June 1, 2013 based on the Farsight study. Dr. Brown refers to the holographic nature of reality and to his remote viewers having possibly viewed a global coastal event happening in June 2013 on Earth in some other timeline or holographic version of Earth or parallel universe (Webre, 2013a).

Planet X and the Catastrophic Timeline

Some had speculated that a Planet X (Nibiru or Second Sun type flyby) would be responsible for a June 2013 global coastal event predicted by the Farsight Institute study (Webre, 2013c).

Dr. Brown's interim update appears to suggest that Dr. Brown did not expect any other meteor or coastal event between February 16, 2013 and June 1, 2013 and that the Farsight Institute report states that no global coastal event caused by Planet X had been expected through June 1, 2013.

Why the Global Coastal Event DARPA Foresaw Did Not Materialize

U.S. chrononaut Andrew D. Basiago writes that the 1971 DARPA probe was actually to an alternative catastrophic time line. He states,

> We are now starting to discover why the global coastal flooding event that I saw when DARPA sent me to 2013 via chronovisor in 1971 didn't eventuate on this time line. NASA is now reporting that in 2012, a solar flare almost destroyed Earth. According to my source to the secret U.S. agency the Department of Physicists, had the huge solar flare occurred six to seven days before it did, when the Earth was one degree of tilt in another position, it would have "wreaked havoc" on Earth. Apparently, the probe I was involved in at ITT Defense Communications in Nutley, NJ on November 5, 1971 for the Office of Naval Intelligence took us to an alternate time line where the devastating effects of such an event in 2012 were evident in 2013, including Washington, D.C. being 100 feet under water. (Webre, 2014a)

Dr. Courtney Brown's February 16, 2013, interim update on the Farsight Institute report and Andrew D. Basiago's conclusions regarding DARPA's 1971 chronovisor probe of 2013 appear both to conclude that the respective catastrophic timelines viewed in each study are not the (non-catastrophic) timelines that Earth is actually experiencing in 2013. (Webre, 2014a)

The False Catastrophic Timeline and False Flag Operations

As shown by time science research (Webre, 2014a), Matrix Elites, Negative dimensional entities, and Matrix institutions (Governmental forms, Banks, and Religions) appear to be operating from (1) duality consciousness ("I win, You lose"), (2) to create a false catastrophic timeline and thus attempt to maintain hegemony over Earth and humanity.

Using the primary playbook of the False Flag Operation, the Matrix powers appear to be employing an End Times script, ecocide, and depopulation through Weather Warfare; Tectonic Warfare; Biowarfare (Ebola; GMOs; Chemtrails; Famine); World War III memes; Negative dimensional warfare; Religious warfare; and other means (Ecologynews.com, n.d.).

These efforts cannot find a stable dimensional reality along the positive timeline and have an increasingly short shelf life in the minds of a humanity that is now bathed in the universe and galactic and solar waves of Unity consciousness.

The Positive Timeline and Unity Consciousness

In 2012-13, Dr. Carl Joseph Calleman, an expert on the Mayan concepts of time and cosmology, stated in an interview with me, "the universal alternating energy wave movements end, and Earth is set on a gradual setting of a potential to reach advanced utopian planet status – a virtual 'Garden of Eden'" (Webre,2011).

Dr. Calleman writes that the positive timeline envisions 2012-13 and the years that follow as: "2012 heralds Earth's entry into the Golden

Age, and between now and then is a time of transition from life as you have known it into life totally in harmony with all of Nature" (Webre,2011).

Universe singularity is now emanating energy for *"enlightened unity consciousness"*.

According to Mayan time and cosmology expert, Dr. Carl Johan Calleman, October 28, 2011, marked a portal in linear time when the singularity [interdimensional portal] at the core of our own universe began to emanate a constant "enlightened universal consciousness" (Webre,2011).

Unity Consciousness is a consciousness that realizes "We are One."

Likewise, the interdimensional portal or singularity at the center of our galaxy (our galactic center "black hole") modulates this universe consciousness energy wave as well.

This is, by any standard, exceedingly good scientific news.

Universal, alternating energy wave movements have been a feature of our universe, says Dr. Carl Johan Calleman, since the Big Bang, and it is these wave movements that have shaped the nature of consciousness in our universe over the past 18 billion years (Webre,2011).

What, you might ask, does this universal wave of unity consciousness have to do with you or your reality?

In a word, everything.

The wave of unity consciousness – like all universe energy wave emanations, creates the "meme" or story content of our personal and our collective reality.

Dr. Calleman's discoveries are suggesting that our highest thoughts and "memes" are/might, in fact, be sourced from the energy waves of the universe singularity, mediated through the galactic singularity – the black hole at the center of our galaxy the Mayans called Hunab Ku (Webre,2011).

The singularity of our solar system is Sol, our sun—a dimensional portal to other galaxies, according to physicist Nassim Haramein. (Haramein, 2012)

These universe singularity energy waves serve as carrier waves of "the universe's mind and spirit software"—the way that the intentional universe (some call it source, the "sea of Light" or God) lets us know of its intention for consciousness in the entire universe.

As a practical matter, does it matter if you are tuned in and commit to Universal consciousness universe energy wave of unity consciousness?

Yes – the universe energy waves we humans tune in and commit to determine our planetary status and how we reach our potential as a planet.

Positive Timeline: Earth will be utopia, not a dystopia.

The good news is that our universe singularity is now emanating "utopia" consciousness waves on its alternating energy carrier waves.

The more we humans are individually and collectively tuned into the universal energy wave of "enlightened universal consciousness," and committed to achieving unity, non-dualistic consciousness, then the stronger the "positive timeline" becomes, and the more rapidly we achieve our potential as a planet – "a potential to reach advanced utopian planet status" (Webre, 2011d) .

The Positive Paradox

The "Positive Paradox" is although Earth is objectively and scientifically on a positive timeline in a field of Unity Consciousness, the more humanity as a critical mass fully internalize and acknowledge this reality, the more it manifests.

The term "conscious" is used in Unity Consciousness to emphasize how important individual commitment to achieving unity consciousness is.

One psychiatrist professor I know stated, "Humans are Velcro for the negative and Teflon for the positive. Because avoiding the negative was so important in our evolution – avoiding getting eaten in the jungle – our brains tend to remember the negative and forget the positive."

Unity consciousness can be expressed in a simple realization: You and I are not we but one.

Dr. Calleman writes,

> To begin with, for all that we know it is designed to bring a shift to unity consciousness where the human mind no longer will be dominated by any dark filter. We will in other words become "transparent" and I believe this is the particular consciousness – seeing reality the way it is with no separation—that so many are waiting for. Not just any consciousness, but one that transcends the dualities of the past and aids the human beings to see the unity of all things.
>
> The reason that this kind of unity consciousness can be beneficial to the planet, and to mankind, is that it is one that leads to the transcendence of all separation (between man and woman, man and nature, ruler and ruled, east and west, etc). I feel that without the manifestation of such a shift in consciousness the world will sooner or later come to an end.
>
> People with a dualist and separating consciousness are somewhat like cancer cells in the body of the Earth with little regard to its larger whole and would eventually generate a collapse of its ecosystem. Only a shift to unity consciousness will forever stop the

unchecked exploitation of the Earth and on a deeper level make us understand that we are part of creation and need to live in harmony with it. But will such a shift just happen automatically? (Webre, 2014a)

Support My Initiative to be a Representative of New Earth on the Regional Galactic Governance Council (Webre, 2018b).

My insight is that fundamental positive transformation on our New Earth will take place when New Earth humanity is consciously and intentionally represented at the regional Galactic Governance Council, so that New Earth can have direct access on policy proposals and petitions for redress for New Earth and New Earth humanity's progress and welfare in our dimensional ecology.

To open the way for creation of this new position in Galactic governance, I ask your support for my initiative to be an Earth's Representative on the Regional Galactic Governance Council.

My Platform as a Representative of New Earth on the Regional Galactic Governance Council is as follows:

1. A full public disclosure and abrogation of all secret treaties with manipulatory extraterrestrial civilizations including and not limited to the following:
    - Draco Reptilian/Orion Grey-USA-Franklin Delano Roosevelt Treaty 1933;
    - Draco Reptilian/Orion Grey-Nazi Germany Treaty-1936-38;
    - Draco Reptilian/Orion Grey-USA-Harry S. Truman Treaty 1948;

- Draco Reptilian/Orion Grey-USA-Dwight D. Eisenhower Greada Treaty 1954;
- Draco Reptilian/Orion Grey-USA-George HW Bush Tau9 Treaty 1989-1993;
- and others.

1. Termination of all Draco reptilian, Orion Gray and any other manipulatory extraterrestrial trafficking, imprisonment, torture, manipulatory experimentation, extraction, consumption, pedocriminal activity of human souls, human beings, human children and infants, human families, human body parts whether pursuant to such Treaties or otherwise.
2. Full termination of presence of manipulatory extraterrestrials on Earth and termination of any and all organizations, institutions, governments, agents of influence, and criminal networks established by such manipulatory networks.
3. Truth and Disclosure—A full public disclosure of the presence of intelligent civilizations in Earth's environment and a global referendum as to whether and on what conditions humanity should enter into relations and space travel, space colonization, and space governance with an organized intelligent universe, a multiverse, or any Omniverse society.
4. A full public disclosure of secret (new energy, zero point, free, antigravity, exotic, etc.) new energy sources now available for application on Earth. Public implemen-

tation and rollout of sequestered free energy technologies for powering dwellings, human settlements, industry, transport and propulsion, communication, and many other energy requirements.

5. Implementation of teleportation as a global, national, regional and local transportation system, replacing polluting fossil fuel vehicles (trains, buses, trucks, autos) and their intensive land use in highways, railways, and urban freeways.
6. Implementation of a regulated time travel public education program.
7. Recognition of Animals as sentient beings with rights—Worldwide grant of personhood rights to animals with concomitant rights against murder, slaughter, torture, and cruel and inhumane treatment.
8. Special intelligent civilization status for cetaceans including whales and dolphins.
9. Development of healthy, safe, tasty protein meat substitutes for humanity's consumption and nutrition.
10. Secure Online Direct Democracy at the local, regional, national, and global level—Secure virtual technology now permits the implementation of Swiss canton democracy worldwide. There is no more need for intermediaries such as City Councils, State or Provincial Legislatures, National Parliaments or Congresses, or even, ultimately in time, a gathering of nations such as the United Nations. Experience over the centuries has

shown that the powers that be buy off all intermediaries. Direct virtual democracy adapts secure virtual technologies and provides virtual hack-proof citizen voting at the municipal, provincial/state, regional, national, and world level. Under direct virtual democracy, the entire city votes on municipal laws; the entire nation votes on national laws; the world population votes on global standards, all duly informed by government staff at the respective local, national, and world level. Municipal Government, for example, is tasked with efficiently picking up the garbage and managing the city according to the laws passed by local virtual democracy.

11. Reinvention of money as a human right and public utility like air, water, or electricity available for creative investment at public money utilities.

12. A global ban on privately controlled central banks such as the U.S. Federal Reserve System and on privately owned commercial banks.

13. Support of complementary currencies.

14. Licensing of consumer cooperatively owned banks.

15. Imposition of heavy criminal penalties for violation and astronomical fines, for individuals, organizations, and nations.

16. Social guarantees in the form of annual income, health care, and elementary, secondary, and post-secondary education for every person on the planet, for life and funded by universal state pools, tax on all financial

transactions and by post graduation contributions to education plan, and more.

17. Implementation of traditional and alternative, as well as advanced extraterrestrial medical technologies.

18. World Debt Forgiveness – Global forgiveness of all public and private debts – a world bankruptcy for a bankrupt system and an end to the debt – fiat money prison system.

19. Criminalization of charging interest on money and of fractional reserve lending.

20. Disenfranchisement of the state power of monarchies and religions worldwide – The UK monarchy and the Vatican are examples of the abuses that occur when two institutions based on non-democratic principles (Divine Right of Kings and Popes) are given established state rights in a modern democratic world.

21. Criminalization of the war industry—A criminalization of and global ban on war, genocide, and depopulation in all its varied forms, overt and covert. A ban on war as a dispute resolution method. A permanent ban on the design, production, or sale of weapons systems, including nuclear weapons, space-based weapons, and conventional weapons. A permanent ban on the maintenance of offensive armed forces. Imposition of heavy criminal penalties for violation and astronomical fines, for individuals, organizations, and nations.

22. Criminal Prosecution and Conviction of a War Crimes

Racketeering Organization and Restorative Justice for War Crimes Victims—Criminalization and rigorous prosecution of the international war crimes racketeering organization for planning and implementing a genocidal depopulation program, including (but not limited to):

a. planning and triggering wars and armed conflicts through false flag operations;

b. regional and global radiation genocide and ecocide through depleted uranium (DU) and the nuclear agenda;

c. planning and implementing environmental war attacks including geo-engineering, weather warfare, HAARP, chemtrails, scalar weapons, robotization and genocide of humanity, famine, vaccines, GMO foods, DNA manipulation and more;

d. Carrying out a program of assassination and Cointelpro terror against activists, researchers, and social inventors in the multiple areas of peace research; new energy; food and nutrition; radiation; democracy, and electoral politics;

e. Carrying out as DOPE INC. a lethal, 300 hundred year old conspiracy to addict humanity to narcotics and to criminalize useful substances such as hemp for profit and enslavement;

f. the transhumanist agenda of population mind control through nano-weapons, EMF and other weapons. There is no statute of limitations on murder.

g. Imposition of heavy criminal penalties for violation and astronomical fines, for individuals, organizations, and nations.

These suggested collective actions and policies for a positive future manifest along humanity's positive timeline.

The basic equation of a positive future suggests that these collective actions and policies (or variations of them to achieve essentially the same goals) will manifest out of the synergy of the positive timeline and humanity's awakening to Unity consciousness.

The Positive Future equation reflects a new level of collective manifestation by humanity and its individuals, resulting from the synergistic dynamics of the positive timeline and a humanity awakening to Unity consciousness.

Who is Futurist Alfred Lambremont Webre?

Futurist Alfred Lambremont Webre's principal social contributions have been (1) founding the science of Exopolitics through his 2000 book *Exopolitics*, (2) the 2014 co-discovery of the Omniverse as the 3rd major cosmological body [after the Universe and the Multiverse] through which humanity understands the cosmos, (3) the development of the Positive Future Equation and the Ascension Hypothesis that describes Soul development in our Universe and Omniverse.

A graduate of Georgetown Preparatory School in classics (1960), Yale University (Industrial Administration Honors-1964), Yale Law School (International Law-1967), University of Texas (Counseling-1997) and a Fulbright Scholar (International Trade-1968), Alfred has taught at two universities (Yale University Economics Department-Taxation) and University of Texas Government Department-Constitutional Law).

Webre has served in leadership positions in environmental protection, international law and justice, public health, exopolitics, and public broadcasting. Webre was General Counsel of the New York City Environmental Protection Administration and later consultant to the Ford Foundation's Public Interest Environmental Law program, overseeing grantees Environmental Defense Fund and Natural Resource Defense Council. Alfred also served as Public Participant, Joint Public Affairs Committee, Commission for Environmental Cooperation (CEC), Montreal, Quebec, Canada.

An international lawyer with Cleary, Gottlieb, Steen and Hamilton, Webre later served as Non-Governmental representative at the United Nations (New York and Vienna) and as Judge on the Kuala Lumpur War Crimes Tribunal, finding the governments of Tony Blair and George W. Bush guilty of war crimes in Iraq. In public health, Alfred served as Deputy Director of the Brownsville Community Health Center, serving 90,000 annual patient encounters in the Lower Rio Grande Valley, earning a Certificate from the Harvard School of Public Health (1995).

In Exopolitics, while a Futurist at Stanford Research Institute, Webre served as director of the 1977 proposed Carter White House Extraterrestrial Communication Study, as well as a Disclosure Project witness in 2001.

In public affairs and politics, Webre served as a Member, Governor's Emergency Taskforce on Earthquake Preparedness, State of California (1980-82); Co-Director, Assassination Information Bureau, Washington, DC, (Public interest counterintelligence to the House Select Committee on Assassinations); Elected Delegate, Texas Democratic Presidential Convention, Dallas, Texas (1996)

A leader in public broadcasting, Webre co-produced the "Instant of Cooperation", the first live radio program in history between the Soviet Union [Russia] and the United States, nominated for an award at the UN General Assembly (1987). Webre has been host on WBAI-FM (New York) and Vancouver Coop Radio and he has been featured on major international networks including CBC-TV, CBS-TV, CNN-TV, TruTV, PressTV, and others.

Leading Books/ Libros Principales:

The Omniverse Tetralogy

Emergence of the Omniverse **(2020: UniverseBooks.com)**

By Alfred Lambremont Webre (Author)

My Journey Landing Heaven on Earth—A Memoir **(2017: UniverseBooks.com)**

By Alfred Lambremont Webre (Author)

"Your autobiography is, in my opinion, the 21st century equivalent of both Augustine of Hippo's Confessions, and The Confessions of Jean-Jacques Rousseau. Your autobiography is revolutionary (as were the autobiographies of those two individuals) in the sense that it has innovated why and how an autobiography is written.

Similar to the two autobiographies mentioned above, in your own autobiography, you have the immense courage to share with the reader both your hardships and your achievements, throughout various phases of your current incarnation; in a *candide* and original manner. Your autobiography stands apart from theirs because you have written it for the altruistic reason of sharing your current "soul journey"; as a star chart, or road map, to help to inform, educate, empower, enlighten, and expand the consciousness of humanity on Earth.

Your autobiography is another "consciousness raising" work, par excellence, Mr. Webre.

Thank you very much for sharing your soul's journey landing Heaven on Earth, thus far! God bless!

READ A SAMPLE and OWN:

My Journey Landing Heaven on Earth

(Webre, 2015g)

http://www.UniverseBooks.com

The Omniverse: Transdimensional Intelligence, Time Travel, the Afterlife, and the Secret Colony on Mars (2015: Inner Traditions/Bear & Co.)

By Alfred Lambremont Webre (Author)

"Although mere mortals can hardly wrap their minds around infinity,

The Omniverse calmly and competently opens the portals to our heavily populated cosmos. A fascinating account of worlds without end, this unusual book illuminates everything from wormhole technology, sky fish, and folds and slots in space to alien abductions, the civilization of souls, and our own government's riveting jump room to Mars (with a young Barack Obama serving as a chrononaut). Here is a preview for Earthlings of the next giant step forward in cosmology and the outer dimensions."

Susan Martinez, PhD, author of Delusions in Science and Spirituality and The Mysterious Origins of Hybrid Man

READ A SAMPLE and OWN:

The Omniverse: Transdimensional Intelligence, Time Travel, the Afterlife, and the Secret Colony on Mars

By Alfred Lambremont Webre

http://www.UniverseBooks.com

Exopolítica: La Política, El Gobierno Y La Ley En El Universo (Vesica Piscis España)

por Alfred Lambremont Webre

Comentario de Paul T. Hellyer Ministro de Defensa Nacional durante el gobierno del primer ministro canadiense Lester B. Pearson; viceprimer ministro de Canadá durante el gobierno del primer ministro Pierre Trudeau:

La odisea de Alfred Lambremont Webre al adentrarse en el reino de la vida que hay en el inmenso Universo que rodea al planeta Tierra, en realidad, es un viaje fascinante si lo lees con una mente abierta. Él postula un Universo en el que existen muchos planetas que sustentan vida y están más avanzados que el nuestro, todos sujetos a la forma de gobierno universal, basada en la regla de la ley.

Él sugiere que la Tierra es una excepción. No sólo no somos el centro del Universo, como creían nuestros antepasados, sino que somos la oveja negra de la comunidad interplanetaria. Fuimos «puestos en cuarentena» y aislados de la «sociedad multidimensional del Universo, una sociedad interplanetaria, intergaláctica y sumamente organizada», supuestamente porque nuestra cultura ha estado fuertemente influenciada por un liderazgo planetario que decidió «ir por libre» y que encontramos personificado en el relato del Jardín del Edén.

Para poner fin a la cuarentena, los terrícolas debemos avanzar ética y espiritualmente, a la vez que restablecemos la conexión con la sociedad interplanetaria. Hasta hace poco tiempo no disponíamos de la tecnología para hacer esto último, pero cada vez tenemos más. Mientras tanto, las visitas de nuestros vecinos extraplanetarios nos brindan oportunidades para entablar una comunicación y colaboración pacíficas.

LEA UNA MUESTRA Y POSEE EL LIBRO ELECTRONICO

Exopolítica: La Política, El Gobierno Y La Ley En El Universo (Vesica Piscis España)

por Alfred Lambremont Webre

http://www.UniverseBooks.com

Exopolitics: Politics, Government, and Law in the Universe (2005: UniverseBooks)

By Alfred Lambremont Webre

The Honorable Paul T. Hellyer, Minister of National Defense under Canadian Prime Minister Lester B. Pearson and Deputy Prime Minister of Canada under Prime Minister Pierre Trudeau states,

Alfred Lambremont Webre's odyssey into the realm of life in the vast Universe surrounding planet Earth is indeed a fascinating journey if you read it with an open mind... To turn us in the direction of re-unification with the rest of creation the author is proposing a "Decade of Contact"—an era of openness, public hearings, publicly funded research, and education about extraterrestrial reality. That could be just the antidote the world needs to end its greed-driven, power-centered madness.

-

Contact/Contacto

U.S. Mailing Address:
936 Peace Portal Drive #39
Blaine, WA 98230 U.S.A

Tel: 604-733-8134
Skype: @peaceinspace
Email: exopolitics@exopolitics.com
NewsInsideOut.com: http://www.NewsInsideOut.com
EXOPOLITICS.com: http://www.exopolitics.com
ExopoliticsTV: http://www.youtube.com/ExopoliticsTV
Facebook: https://www.facebook.com/alfred.webre
Twitter: @alfredwebre
2020:
Omniversity.net
http://www.omniversity.net
Omniversidad.info
http://www.omniversidad.info

References

- "Ben 10: Omniverse (Video Game)" (n.d.), *Wikipedia*, Wikimedia Foundation, 4 Aug. 2019, https://en.wikipedia.org/wiki/Ben_10:_Omniverse_(video_game)
- "Mars Jump Room with Mars Explorers Andrew D. Basiago, Bernard Mendez, William B. Stillings & Alfred Lambremont Webre [Moderator] "(n.d.), *ExoSciences Curriculum, Omniversity.net,*
- Alvarado, C.S. (2017), "Ernesto Bozzano's "Phénomènes Psychiques au Moment de la Mort" (Psychic Phenomena at the Moment of Death)", *Psi Encyclopedia,* psi-encyclopedia.spr.ac.uk/articles/ernesto-bozzanos-phenomenes-psychiques-au-moment-de-la-mort-psychic-phenomena-moment-death
- Ascension Cycle, Precession of the Equinoxes (n.d.), *Ascension Glossary* https://ascensionglossary.com/index.php/Precession_of_the_Equinoxes
- *Amazon*, Amazon, 1 Jan. 1971, Retrieved from www.amazon.com/Psychic-Discoveries-Behind-Iron-Curtain/dp/0553135961 .
- Belitsos, Byron, The end of the Lucifer rebellion and the coming of the Avatar of our Age, http://exopolitics.blogs.com/exopolitics/2013/12/byron-belitsos-the-end-of-the-lucifer-rebellion-and-the-coming-of-the-avatar-of-the-age.html
- Bertolacci, D. (2014), "Grand Slam Theory of the Omniverse."*BalboaPress*, BalboaPress, 20 Aug. 2014, www.balboapress.com/bookstore/bookdetail.aspx?bookid=SKU-000622464.
- Branton, A. (2015), The Dulce Book, Retrieved from all-natural.com/wp-content/uploads/2015/04/Dulce_Book_-_Area_51_Underground.pdf
- Branton, A. (2016), Dulce Base Security Officer Speaks Out-The Dulce Book – Chapter 11 The Prisoners of Dulce Base Retrieved from www.auricmedia.net/wp-content/uploads/2016/04/The-Prisoners-of-Dulce-Base.

pdf all-natural.com/dulce-11/

- Braude, S. E. (2016), "Postmortem Survival". *Psi Encyclopedia.* psi-encyclopedia.spr.ac.uk/articles/postmortem-survival .
- Cameron, G. (2017), "Portals and Xendras: a 2014 Story." *Earth Mystery News*, 10 Oct. 2017, Earthmysterynews.com/2017/10/10/portals-and-xendras-a-2014-story/ .
- Cook, E. W., Greyson, B.[sic], & Stevenson, I. (n.d.), "Do Any Near-Death Experiences Provide Evidence for the Survival of Human Personality after Death?" Relevant Features and Illustrative Case Reports, med.virginia.edu/perceptualstudies/wpcontent/uploads/sites/360/2017/01/STE46_Do-Near-Death-Experiences-Provide-Evidence-for-Survival-of-Human-Personality.pdf
- Ecologynews.com (n.d.). Please see extensive documentation in subject articles at Ecologynews.com.
- *Ever Beyond Radio* (2015), 178.62.57.183/?p=1706
- Exopolitics.com (2017), exopolitics.blogs.com/exopolitics/2017/06/exopolitical-drivers-of-pedocriminal-networks-abrogate-renegotiate-secret-greada-tau-9-treaties-with-pedocriminal-ets-dr.html
- Freeland, E. (2014) , Chemtrails, HAARP, and the Full Spectrum Dominance of Planet Earthhttps://www.amazon.com/Chemtrails-HAARP.../dp/1936239930
- Freeman, M. (2015), "Soul-Catching Net—Are We 'Recycled' at Death to Remain in the Matrix?" *Wake Up World*, 13 Mar. 2017, wake-up-world.com/2015/07/23/soul-catching-net-are-we-recycled-at-death-to-remain-in-the-matrix/.
- Goswami, A. "The Idealistic Interpretation of Quantum Mechanics." Phys. Essays 2 (1989): 385–400. ———. "Consciousness in Quantum Physics and the Mind-body Problem." J. Mind and Behavior 11, no. 1 (Winter 1990): 75–96. ———. The Self-Aware Universe: How Consciousness Creates the Material World. New York: Tarcher/Putnam, (1993), "Monistic Idealism May Provide Better Ontology for Cognitive Science: A Reply to Dyer." J. Mind and Behavior 16, no. 2 (Spring 1995: 135-50.)

- Greyson, B. (2017a), Cosmological Implications of Near-Death Experiences, med.virginia.edu/perceptualstudies/wpcontent/uploads/sites/360/2017/01/NDE65.pdf
- Greyson, B. (2017b), Implications of Near-Death Experiences for a Postmaterialist Psychology, med.virginia.edu/perceptualstudies/wpcontent/uploads/sites/360/2017/01/NDE62_postmaterialist-PRS.pdf
- Haramein, N. (2012), The Sun is a Stargate, https://truedemocracyparty.net/2012/06/stargate-sun-the-sun-is-a-stargate-all-stars-are-stargates-highly-advanced-extraterrestrial-technology-opens-gateway-into-our-solar-system-some-nasa-censorship-involved/
- Hansen, & Schild, (n.d.), The Dual Soul Connection: The Alien Agenda for Human Advancement, Paperback and Kindle eBook: www.communicatorlink.com/
- Haraldsson, E. (2017), "Deathbed Visions Research", *Psi Encyclopedia*, psi-encyclopedia.spr.ac.uk/articles/deathbed-visions-research
- https://www.bibliotecapleyades.net/ciencia/ciencia_consciousuniverse20.htm
- Infiltrating the New Age Movement (n.d.), *Ascension Glossary*ascensionglossary.com/index.php/Infiltrating_the_New_Age_Movement
- Ipsos MORI (n.d.), "People in western countries pessimistic about future for young people", Ipsos MORI Global Trends Survey ipsosmori.com/researchpublications/researcharchive/3369/People-in-western-countries-pessimistic-about-future-for-young-people.aspx
- Kautz-Vela, H. (n.d.),, "Silent Assimilation: A.I. Black Goo, Control of the Human System", *Off-Planet TV,* https://youtube.com/watch?t=201&v=MaP5OAcg3g8
- Kelly, E. W. (2017), "Near-Death Experiences With Reports Of Meeting Deceased People", med.virginia.edu/perceptualstudies/wpcontent/uploads/sites/360/2017/01/KEL13-NDEwithReports-of-Meeting-Deceased-People.pdf
- Kelly, E. W. PhD, Greysun, B. MD, & Stevenson, I. MD (1999-2000), "Can Experiences Near Death Furnish Evidence of Life After Death?"

med.virginia.edu/perceptualstudies/wpcontent/uploads/sites/360/2017/01/STE50-Omega-1999-2000-NDE-paper.pdf

- Kolosowa, L. E. (n.d.), "Earth Timeline/Matrix Timeline (plan) P3", www.youtube.com/watch?v=ZL6LM5cZhc&list=PL0H2nHL6XjmdhD4Fnad-kNcdNLsAH9dxoL&index=4
- Lazlo, E. (n.d.), "An unexplored domain of nonlocality: Toward a scientific explanation of Instrumental Transcommunication" www.sciencedirect.com/science/article/pii/S1550830708002000
- Linde, A. & Vanchurin, V. (2015), "How Many Universes Are in the Multiverse?" http://arxiv.org/pdf/0910.1589v3.pdf
- Long, J. MD (n.d.), "Evidence of the Afterlife: The Science of Near-Death Experiences", (New York: Harper Collins) www.amazon.com/Evidence-Afterlife-Science-Near-Death-Experiences-ebook/dp/B0032JQ7D0/ref=sr_1_7?s=digital-text&ie=UTF8&qid=1545611930&sr=17&keywords=near+death+experiences
- Loves, B. (2015), "Artificial Intelligence & Deep Mind Control" https://bradleyloves.wordpress.com/2015/08/08/artificial-intelligence-a-i-part-one/
- Lucid-Mind-Center.com (n.d.), Proof Of Life After Death, *Robert Monroe's "Far Journeys",* www.lucid-mind-center.com/proof-of-life-after-death.html
- Maddaloni, V. (1972), "Invention of a Machine for Photographing the Past." *Domenica del Corriere 74, no.18* (Might 2, 1972).
- Miller, J. S. (2012), Near-Death Experiences As Evidence for the Existence of God and Heaven *(Georgia: 2012 Wisdom Creek Press), Chapter 2.* www.amazon.com/Near-Death-Experiences-Evidence-Existence-Heaven-ebook/dp/B00A3MU6H2/ref=sr_1_19?s=digital-text&ie=UTF8&qid=1545614598&sr=119&keywords=near+death+experiences
- Morgan, E. (2018), "Harald Kautz Vella; Morgellons & Smart Dust Infect Individuals to be Tracked via Cell Towers" February 13, 2018 https://prepareforchange.net/2018/02/13/harald-kautz-vella-morgellons-smart-dust-infect-individuals-to-be-tracked-via-cell-towers/
- Newton, M. (2008), Journey of Souls: Case Studies of Life Be-

tween Lives. Woodbury, Minn.: Llewellyn, 2008.

- Newton, M. (2009), Destiny of Souls: New Case Studies of Life Between Lives. Woodbury, Minn.: Llewellyn, 2009.
- Omnisense/Walker, P. (2010), "Omniverse." December 24, 2010, *Urban Dictionary*, www.urbandictionary.com/define.php?term=Omniverse
- Ostrander, S., & Schroeder, L. (1971), "Psychic Discoveries Behind the Iron Curtain."
- Panel (n.d.), "AI & Mind Control: Parasitic Consciousness, Multi Level Mind Control, And The Transhumanist Agenda". *CCN-TV* youtube.com/watch?v=50kE55EvRw8
- Parry, R. (2013), "A CIA Hand in an American 'Coup'?" *Global Research*, 26 Aug. 2013, www.globalresearch.ca/a-cia-hand-in-an-american-coup/5346984.
- Penre, W. (2011), "Metaphysics Paper #4: There is a Light at the End of the Tunnel—What Happens After Body Death?" *Steemit*, https://steemit.com/spiritual/@titanblooded/what-happens-after-body-death . www.scribd.com/document/96096852/MetaphysicsPaper4-There-is-a-Light-at-the-End-of-the-Tunnel-What-Happens-After-Body-Death
- Penre, W. (2014), Definitions of Special Terminology, First-Fifth Levels of Learning, *Wes Penre Publications,* www.bibliotecapleyades.net/ciencia/historia_humanidad56.htm
- Penre, W. (2016), Synthetic Super Intelligence and the Transmutation of Humankind, 2016, Wes Penre Publications *https://wespenre-*publicationshome.files.wordpress.com/2019/02/wes_penre___synthetic_super_intelligence_and_the_transmutation_of_man__a_roadmap_to_the_singularity.pdf
- Penre, W. (2019a), "Article#2: The Death Trap and How to Avoid It." Wes Penre Publications, Retrieved 17 Feb. 2019 from https://wespenrepublications.home.blog/2018/12/18/article-2-the-death-trap-and-how-to avoid-it/
- Penre, W. (2019b), "III. The Reincarnation System is being Refined! (Additional Ideas on How the Afterlife is Set Up)", Wes Penre Publi-

cations https://wespenre.com/2019/02/04/fourth-level-of-learning-paper-15-the-postdiluvian-times-and-lucifer-building-his-plans/

- Penre, W. (2019c), "Article #5: Why We Need to Get Rid of Attachments to Exit This Matrix." *Wes Penre Publications*, 2 July 2019, https://wespenrepublications.home.blog/2019/01/07/article-5-why-we-need-to-get-rid-of-attachment-to-exit-this-matrix/
- Penre, W. (2019d), Fifth Level of Learning, Paper 1: Hindu Cosmology, *Wes Penre Publications,https://*wespenre.com/2019/02/05/fifth-level-of-learning-paper-1-hindu-cosmology/
- Penre, W. (2019e), Synthetic Super Intelligence and the Transmutation of Humankind, pages 406-419, https://wespenrepublicationshome.files.wordpress.com/2019/02/wes_penre___synthetic_super_intelligence_and_the_transmutation_of_man__a_roadmap_to_the_singularity.pdf.
- Penre, W. (n.d.a), "Metaphysics Paper #4: There is a Light at the End of the Tunnel—What Happens After Body Death?" *Steemit*, https://steemit.com/spiritual/@titanblooded/what-happens-after-body-death . www.scribd.com/document/96096852/MetaphysicsPaper4-There-is-a-Light-at-the-End-of-the-Tunnel-What-Happens-After-Body-Death
- Renee, L. (n.d.), "Transhumanism & Artificial Intelligence" isabeaux.net/news/display.php?M=107919&C=ccb6091f95f940d825c115dc53592a84&S=494&L=14&N=308
- Sainato, M. (2015), "Stephen Hawking, Elon Musk, and Bill Gates Warn About Artificial Intelligence" observer.com/2015/08/stephen-hawking-elon-musk-and-bill-gates-warn-about-artificial-intelligence/
- Sartori, P. (2015). "Near-Death Experience". *Psi Encyclopedia,* psi-encyclopedia.spr.ac.uk/articles/near-death-experience
- Scheck, R., & Piacenza G. (2019), "Bassett on TTSA – The Most Important Development In History of Disclosure Movement! ." *EXONEWS*, 30 Apr. 2019, exonews.org/tag/extraterrestrial-disclosure/.
- Senkowski, E. (n.d.), "Instrumental Transcommunication ITC—in short," www.worlditc.org
- Stevenson, I., MD (n.d.), "Reincarnation Research" www.

neardeath.com/reincarnation/research/ianstevenson.html#a02

- Swedenborg Foundation (n.d.), Spiritual World (Afterlife), swedenborg.com/emanuel-swedenborg/explore/spiritual-world/
- Swedenborg, E. (n.d.), *Heaven and Hell*, sections #445–52 swedenborg.com/emanuel-swedenborg/writings/short-excerpts-and-downloads/entry-into-eternal-life/
- Technological singularity (n.d.), en.wikipedia.org/wiki/Technological_singularity
- The Ruiner (2015), "Earth Based Artificial Intelligence – A.I.", August 12, 2015, bradleyloves.wordpress.com/2015/08/12/Earth-based-artificial-intelligence-a-i/
- The Urantia Book (1955), Paper 53, The Lucifer Rebellion *(1955: The Urantia Foundation)* www.urantia.org/urantia-book-standardized/paper-53-lucifer-rebellion
- Timeloop Consortium (n.d.), timeloopsolution.com/english/index_e.html
- TOLEC, Email to Alfred Lambremont Webre, September 17, 2019
- University of Virginia Medical School's Division of Perceptual Studies (n.d.) , "Near-Death Experiences" [NDEs] med.virginia.edu/perceptual-studies/research-area/near-death-experiences-ndes/
- Webre, A. L. & Enoch, J. (n.d.), "Exopolitics & The Dimensional Ecology of the Omniverse, *Omniversity* #1, youtube.com/watch?v=XYgMoyYGz4c
- Webre, A. L. (1977), "Proposed 1977 Carter Extraterrestrial Communication Study." EXOPOLITICS, exopolitics.blogs.com/exopolitics/2007/05/proposed_1977_c.html.
- Webre, A. L. (2001a), "Affidavit Of Alfred Lambremont Webre, Filed at Might 9, 2001 Disclosure Project Press Conference, Washington, DC." *Exopolitics*, exopolitics.blogs.com/exopolitics/2017/09/affidavit-of-alfred-lambremont-webre-filed-at-might-9-2001-disclosure-project-press-conference-washing.html.
- Webre, A. L. (2001b),"i disastri globali che stanno colpendo il nos-

tro pianeta, sono gli effetti della "Kill Zone" di Planet X" | Segni dal Cielo—Portale web di UFO News, Cerchi nel grano, profezie mighta, Convegni e seminari segnidalcielo.it/alfred-webre-i-disastri-globali-che-stanno-colpendo-il-nostro-pianeta-sono-gli-effetti-della-kill-zone-di-planet-x/

- Webre, A. L. (2005), "The 1977 Carter White House Extraterrestrial Communication Study", exopolitics.blogs.com/exopolitics/2005/01/the_1977_carter.html
- Webre, A. L. (2010a), "Is the ET Council 2015 ecology clean-up a prelude to 2025 'Paradise on Earth'?" exopolitics.blogs.com/exopolitics/2010/12/is-the-et-council-2015-ecology-cleanup-a- prelude-to-2025-paradise-on-Earth.html
- Webre, A. L. (2010b), "NOTICIAS EXOPOLITICAS – Consejo de 8 Extraterrestre: 'Vamos a aumentar los avistamientos OVNIs, hablar en las Naciones Unidas en 2014, y limpiar el ambiente de la Tierra en el año 2015' STANLEY.A.FULHAM" exopolitics.blogs.com/exopolitics/2010/10/noticias-exopoliticas-consejo-de-8- extraterrestre-vamos-a-aumentar-los-avistamientos-ovnis-hablar-en.html
- Webre, A. L. (2010c), "Are the UFOs appearing in Russia skies in Dec. 2010 Fulham's predicted UFO wave?", exopolitics.blogs.com/exopolitics/2010/12/are-the-ufos-appearing-in-russia-skies-in-dec-2010-fulhams-predicted-ufo-wave.html
- Webre, A. L. (2010d), "Exopolitics founder: 'Oct 13 2010 NYC UFO sightings confirm Exopolitics model", exopolitics.blogs.com/exopolitics/2010/10/exopolitics-founder-oct-13-2010-nyc-ufo-sightings-confirm-exopolitics-model.html
- Webre, A. L. (2010f), "Video: Challenges of Change interview with former NORAD officer Stanley A. Fulham (authentic interview copy)" exopolitics.blogs.com/exopolitics/2010/12/video-challenges-of-change-interview-with-former-norad-officer-stanley-a-fulham-authentic-interview-.html
- Webre, A.L.(2010g),"Universe Singularity Now Emanating Pre-Wave Energy for Enlightened Unity Consciousness" (July 20, 2010)
- Webre, A. L. (2011a), "Israeli media: 'Jerusalem UFO orbs ful-

fill Stan Fulham's ET council prediction'", exopolitics.blogs.com/exopolitics/2011/02/israeli-media-jerusalem-ufo-orbs-fulfill-stan- fulhams-et-council-prediction.html

- Webre, A. L. (2011b), "OVNIS en Rusia: Stanley Fulham who Art in Heaven. Alfred Webre: UFOs Russia (video de Roberto Benitez)" exopolitics.blogs.com/exopolitics/2011/01/ovnis-en-rusia-stanley-fulham-who-art-in- heaven-alfred-webre-ufos-russia-video-de-roberto-benitez.html
- Webre, A. L. (2011d), "Good News! Universe singularity now emanating pre-wave energy for 'enlightened unity consciousness'" exopolitics.blogs.com/exopolitics/2011/12/good-news-universe-singularity-now-emanating-pre-wave-energy-for-enlightened-unity-consciousness.html
- Webre, A. L. (2011g), "Time travel & teleportation", *Articles,* exopolitics.blogs.com/exopolitics/2011/05/quantum-access-time-travel-teleportation-articles-by-alfred-lambremont-webre.html
- Webre, A. L. (2011h), "UFO Expert: Stanley Fulham UFO predictions fulfilled over NY, Moscow, and London", exopolitics.blogs.com/exopolitics/2011/02/ufo-expert-stanley-fulham-ufo-predictions-fulfilled-over-ny-moscow-and-london.html
- Webre, A. L. (2012a), "Are you on a 2012-13 catastrophic or positive future timeline? Part 1", exopolitics.blogs.com/exopolitics/2011/12/are-you-on-a-2012-13-catastrophic-or-positive-future-timeline-part-i.html
- Webre, A. L. (2012b), "Hermeneutics expert: Obama will bring "peace" and ET false flag attack followed by an "Armageddon" against good ETs", exopolitics.blogs.com/exopolitics/2012/04/hermeneutics-expert-obama-will-bring-peace-and-et-false-flag-attack-followed-by-an-armageddon-agains.html
- Webre, A. L. (2012c), "'Occupy Adam's Calendar' Part I Extraterrestrial Genetic Manipulation: Geneticist William Brown—A Film by Alfred Lambremont Webre." *Exopolitics*, exopolitics.blogs.com/exopolitics/2012/06/occupy-adams-calendar-part-i-genetic-manipulation-geneticist-william-brown-a-film-by-alfred-lambremo.html
- Webre, A. L. (2013a), "Dr. Courtney Brown—Russia meteor &

asteroid 2012 DA14 might constitute 2013 "global coastal event" remote viewed in another timeline", exopolitics.blogs.com/exopolitics/2013/02/video-dr-courtney-brown-russia-meteor-asteroid-2012-da14-might-constitute-2013-global-coastal-event-re.html

• Webre, A. L. (2013b), "Mary Rodwell: ETs, Souls, The New Humans, and a Coming Global Shift." *YouTube*, 3 July 2013, youtu.be/MPRN-OkuaYio.

• Webre, A. L. (2013c), "Planet X Update: Is Earth on a positive timeline, 2013-2020?", exopolitics.blogs.com/exopolitics/2013/02/is-Earth-on-a-positive-timeline-2013-2020.html

• Webre, A. L. (2014a), "Creating a Positive Future: Time Science Shows Our Earth is on a Positive Timeline in our Time Space Hologram", exopolitics.blogs.com/positive_future/2014/09/creating-a-positive-future-time-science-shows-our-Earth-is-on-a-positive-timeline-in-our-time-space-hologram.html#more

• Webre, A. L. (2014b), The Dimensional Ecology of the Omniverse (2014 Universe Books) https://www.amazon.com/gp/product/097376631X/ref=dbs_a_def_rwt_bibl_vppi_i.

• Webre, A. L. (2015a), The Omniverse Transdimensional Intelligence, Time Travel, the Afterlife, and the Secret Colony on Mars, (Vermont: 2015 Bear & Co.) www.simonandschuster.ca/books/The-Omniverse/Alfred-Lambremont- Webre/9781591432166

• Webre, A. L. (2015b), "Suzanne Hansen: Dual Human-Grey Soul Earth Education mission by benevolent Grey ETs" newsinsideout.com/2015/11/suzanne-hansen-dual-human-grey-soul-Earth-education-mission-by-benevolent-grey-ets/

• Webre, A. L. (2015c), "Off-planet Artificial Intelligence AI is mobilizing in 2015 for planetary takeover. AI singularity in 2045 is an AI deception", newsinsideout.com/2015/08/off-planet-artificial-intelligence-ai-is-mobilizing-in-2015-for-planetary-takeover-ai-singularity-in-2045-is-an-ai-deception/

• Webre, A. L. (2015d), "Panel finds *prima facie* evidence for sentient, inorganic AI Artificial Intelligence & its stealth takeover of living

Earth and humanity" newsinsideout.com/2015/08/panel-finds-prima-facie-evidence-for-sentient-inorganic-ai-artificial-intelligence-its-stealth-takeover-of-living-Earth-and-humanity/

- Webre, A. L. (2015f),. "Is the ET Council 2015 ecology cleanup a prelude to 2025 'Paradise on Earth'?" exopolitics.blogs.com/exopolitics/2010/12/is-the-et-council-2015-ecology-cleanup-a-prelude-to-2025-paradise-on-Earth.html
- Webre, A. L. (2015g),. "My Journey Landing Heaven on Earth by Alfred Lambremont Webre." *Goodreads*, Goodreads, 24 June 2015, www.goodreads.com/book/show/25808220-my-journey-landing-heaven-on-Earth.
- Webre, A. L. (2016b), "Evidence DARPA-CIA Time Travel Pre-Identified Trump as Future U.S. President." *NewsInsideOut*, 9 Nov. 2016, newsinsideout.com/2016/11/evidence-darpa-cia-time-travel-pre-identified-trump-future-u-s-president/.
- Webre, A. L. (2016c), "The Adjudication of the Lucifer Rebellion Is Done and Planetary Transformation Is Underway." NewsInsideOut, 7 Aug. 2016, newsinsideout.com/2015/07/the-adjudication-of-the-lucifer-rebellion-is-done-and-planetary-transformation-is-underway/
- Webre, A. L. (2017b), "Is a January 28, 2011 UFO Orb over Dome of the Rock in Jerusalem a Context Communication by Interdimensional ET?" exopolitics.blogs.com/exopolitics/2017/12/is-ufo-orb-over-dome-of-the-rock-in-jerusalem-a-context-communication-by-interdimensional-et.html
- Webre, A. L. (2017c), "Patricia Cori-The New Sirian Revelations – Galactic Prophecies for the Ascending Human Collective", newsinsideout.com/2017/12/webinar-patricia-cori-new-syrian-revelations-galactic-prophecies-ascending-human-collective/ www.northatlanticbooks.com/shop/the-new-sirian-revelations/
- Webre, A. L. (2017d), "Download Free PDF Special Report- The UFO/ET Abductions of Jimmy Carter, Alfred Lambremont Webre & Andrew D. Basiago." *EXOPOLITICS*, exopolitics.blogs.com/exopolitics/2017/09/download-free-46-page-pdf-.html.

- Webre, A. L. (2017e),"ET Regional Galactic Governance Council: We will decloak our Spaceship fleet; Speak at the United Nations; and Renew the Earth's Ecology", xopolitics.blogs.com/exopolitics/2017/11/et-regional-galactic-governance-council-we-will-decloak-our-spaceship-fleet-speak-at-the-united-nati.html
- Webre, A. L. (2017f),"Exopolitical Drivers of Pedocriminal Networks: Abrogate & renegotiate secret Greada, Tau-9 Treaties with pedocriminal ETs: Draco reptilians, Orion Greys & Anunnaki ETs" newsinsideout.com/2017/05/exopolitical-drivers-pedocriminal-networks-abrogate-renegotiate-secret-greada-tau-9-treaties-pedocriminal-ets-draco-reptilians-orion-greys-anunnaki-ets/
- Webre, A.L. (2017g), WEBINAR: World Oneness Day—United Nations Day (October 24, 2017)—falls at the maximum peak of the Ninth Wave: Dr. Carl Johan Calleman, Author of THE NINE WAVES OF CREATION.
- https://newsinsideout.com/2017/10/webinar-world-oneness-day-united-nations-day-october-24-2017-falls-maximum-peak-ninth-wave-dr-carl-johan-calleman-author-nine-waves-creation/
- Webre, A. L. (2018a), "Peter Kling: 2020-21 End of Religion; 3.5 years of 666 Prince William; Nibiru & System of Evils falls—Christ Consciousness lands", newsinsideout.com/2018/08/peter-kling-2020-21-end-of-religion-exposes-vatican-planetary-child-trafficking-empire-and-islam-as-a-death-cult-followed-by-3-5-years-of-666-antichrist-king-of-the-world-prince-william-as-a-nibiru/
- Webre, A. L. (2018b), "Support My Initiative to be Earth's Representative on the Regional Galactic Governance Council", Retrieved from https://exopolitics.blogs.com/positive_future/2018/03/support-my-initiative-to-be-Earths-representative-on-the-regional-galactic-governance-council.html
- Webre, A. L. (2019a), "AI Sentient Inorganic Invading Artificial Intelligence", *Articles, Interviews & Independent Panels – 2019* exopolitics.blogs.com/exopolitics/2019/06/ai-sentient-inorganic-invading-artificial-intel-

ligence-articles-interviews-by-alfred-lambremont-webre.html

- Webre, A. L. (2019b),"ET Space craft hovering over July 20, 2019 Pine Ridge All Nations Lakota Sundance fulfills Xico Xavier 50-year prophecy: No WWIII & Humanity will join peaceful Galactic Community: Mia Feroleto" newsinsideout.com/2019/07/et-craft-over-july-20-2019-lakota-sundance-fulfills-xico-xavier-50-year-prophecy-no-wwiii-humanity-will-join-galaxy/
- Webre, A. L. (2019c),"Xico Xavier, Lakota Sundance, Consciousness-Contact, ET-UFOs: No WWIII & Earth Humanity joins Omniverse of Galactic Nations with Ananda Bosman", newsinsideout.com/2019/08/xico-xavier-lakota-sundance-consciousness-contact-et-ufos-no-wwiii-Earth-humanity-joins-omniverse-of-galactic-nations-with-ananda-bosman/
- Webre, A. L., JD & Liss, P. H. PhD (1974), The Age Of Cataclysm, (1974-New York: GP Putnam's Sons, Berkley Paperback) exopolitics.blogs.com/exopolitics/2012/07/the-age-of-cataclysm-prophetic-1974-book-now-required-reading-about-predicted-impacts-of-a-2013-brow.html
- Wehrstein, K. (2017a), "International Reincarnation Cases", *Psi Encyclopedia,* psi-encyclopedia.spr.ac.uk/articles/international-reincarnation-cases
- Wehrstein, K. (2017b), "Pam Reynolds (Near-Death Experience)", *Psi Encyclopedia,* psi-encyclopedia.spr.ac.uk/articles/pam-reynolds-near-death-experience
- Wehrstein, K. (2018a), "AWARE NDE Study", *Psi Encyclopedia*, psi-encyclopedia.spr.ac.uk/articles/aware-nde-study
- Wehrstein, K. (2018b). "Robert Monroe". *Psi Encyclopedia*, psi-encyclopedia.spr.ac.uk/articles/robert-monroe
- Wehrstein, K.(2018c), "Eben Alexander", *Psi Encyclopedia,* psi-encyclopedia.spr.ac.uk/articles/eben-alexander
- World Death Clock (n.d.), www.medindia.net/patients/calculators/world-death-clock.asp
- www.northatlanticbooks.com/shop/the-new-sirian-revelations/
- www.thewatcherfiles.com/dulce/chapter33.htm (30 of 46)8/14/2004 3:27:20 PM

- Zaghmut Wise, E. (2011; 2013), "The Omniverse Defined." *Pantheism Today*, 1 Jan. 1970, pantheismtoday.blogspot.com/2013/01/the-omniverse-defined.html
- Zammit, V. (n.d.), A Lawyer Presents the Case for the Afterlife Irrefutable Objective Evidence, www.victorzammit.com/book/4thedition/chapter29.html
- Ziewe, J. (n.d.a), "Vistas of Infinity—How to Enjoy Life When You Are Dead." www.lulu.com/shop/http://www.lulu.com/shop/jurgen-ziewe/vistas-of-infinity-how-to-enjoy-life-when-you-are-dead/ebook/product-23686922.html
- Ziewe, J.(n.d.b), Communicating with Dead People During OBEs, www.afterlifestudies.org/communicating-dead-people-obes/

P.50 Omniverse – all universes (multiverse) + the spiritual dimensions

P.60 – THE REAL CONSPIRACY IS ABOUT SUPRESSING YOUR IDEA OF WHO YOU REALLY ARE

P.88 – archons

P.143 – Matrix

P.175 – Keep it simple

P.179 – The grid now has holes in it

~~P.185~~
P.185 → PPAI

P.248 – Hologram is made by Source (God)

P.69 – Universes as bubbles

CPSIA information can be obtained
at www.ICGtesting.com
Printed in the USA
LVHW051049230420
654317LV00006B/622